Learn, Practice, Succeed

Eureka Math®

Grade 8
Module 4

Published by Great Minds®.

Printed in the U.S.A.

This book may be purchased from the publisher at eureka-math.org.
10 9 8 7 6 5 4 3 2

ISBN 978-1-64054-983-8

G8-M4-LPS-05.2019

Students, families, and educators:

Thank you for being part of the *Eureka Math®* community, where we celebrate the joy, wonder, and thrill of mathematics.

In *Eureka Math* classrooms, learning is activated through rich experiences and dialogue. That new knowledge is best retained when it is reinforced with intentional practice. *The Learn, Practice, Succeed* book puts in students' hands the problem sets and fluency exercises they need to express and consolidate their classroom learning and master grade-level mathematics. Once students learn and practice, they know they can succeed.

What is in the Learn, Practice, Succeed *book?*

Fluency Practice: Our printed fluency activities utilize the format we call a Sprint. Instead of rote recall, Sprints use patterns across a sequence of problems to engage students in reasoning and to reinforce number sense while building speed and accuracy. Sprints are inherently differentiated, with problems building from simple to complex. The tempo of the Sprint provides a low-stakes adrenaline boost that increases memory and automaticity.

Classwork: A carefully sequenced set of examples, exercises, and reflection questions support students' in-class experiences and dialogue. Having classwork preprinted makes efficient use of class time and provides a written record that students can refer to later.

Exit Tickets: Students show teachers what they know through their work on the daily Exit Ticket. This check for understanding provides teachers with valuable real-time evidence of the efficacy of that day's instruction, giving critical insight into where to focus next.

Homework Helpers and Problem Sets: The daily Problem Set gives students additional and varied practice and can be used as differentiated practice or homework. A set of worked examples, Homework Helpers, support students' work on the Problem Set by illustrating the modeling and reasoning the curriculum uses to build understanding of the concepts the lesson addresses.

Homework Helpers and Problem Sets from prior grades or modules can be leveraged to build foundational skills. When coupled with *Affirm®, Eureka Math*'s digital assessment system, these Problem Sets enable educators to give targeted practice and to assess student progress. Alignment with the mathematical models and language used across *Eureka Math* ensures that students notice the connections and relevance to their daily instruction, whether they are working on foundational skills or getting extra practice on the current topic.

Where can I learn more about Eureka Math *resources?*

The Great Minds® team is committed to supporting students, families, and educators with an evergrowing library of resources, available at eureka-math.org. The website also offers inspiring stories of success in the *Eureka Math* community. Share your insights and accomplishments with fellow users by becoming a *Eureka Math* Champion.

Best wishes for a year filled with "aha" moments!

Jill Diniz

Jill Diniz
Chief Academic Officer, Mathematics
Great Minds

Contents

Module 4: Linear Equations

Topic D: Systems of Linear Equations and Their Solutions

Topic E (Optional): Pythagorean Theorem

Exercises

Write each of the following statements using symbolic language.

1. The sum of four consecutive even integers is -28.

2. A number is four times larger than the square of half the number.

3. Steven has some money. If he spends $\$9.00$, then he will have $\frac{3}{5}$ of the amount he started with.

4. The sum of a number squared and three less than twice the number is 129.

5. Miriam read a book with an unknown number of pages. The first week, she read five less than $\frac{1}{3}$ of the pages. The second week, she read 171 more pages and finished the book. Write an equation that represents the total number of pages in the book.

Lesson Summary

Begin all word problems by defining your variables. State clearly what you want each symbol to represent.

Written mathematical statements can be represented as more than one correct symbolic statement.

Break complicated problems into smaller parts, or try working them with simpler numbers.

Name ______________________________ Date ______________

Write each of the following statements using symbolic language.

1. When you square five times a number, you get three more than the number.

2. Monica had some cookies. She gave seven to her sister. Then, she divided the remainder into two halves, and she still had five cookies left.

1. George is four years older than his sister Sylvia. George's other sister is five years younger than Sylvia. The sum of all of their ages is 68 years.

   Let x be Sylvia's age. Then,
   $(x+4)+(x-5)+x=68.$

Since I know something about Sylvia and both her brother and sister, I will define my variable as Sylvia's age.

2. The sum of three consecutive integers is 843.

   Let x be the first integer. Then,
   $x+(x+1)+(x+2)=843.$

I remember that consecutive means one after the next. If my first number was 5, then a numeric statement would look like this:

$5+(5+1)+(5+2).$

I need to write something similar using symbols.

3. One number is two more than another number. The sum of their squares is 33.

   Let x be the smaller number. Then,
   $x^2+(x+2)^2=33.$

4. When you add 42 to $\frac{1}{3}$ of a number, you get the number itself.

   Let x be the number. Then,
   $\frac{1}{3}x+42=x.$

I don't know what number a fraction of 45 is. I remember that taken away from 23 means I need to subtract the number from 23.

5. When a fraction of 45 is taken away from 23, what remains exceeds one-half of eleven by twelve.

   Let x be the fraction of $\mathbf{45}$***. Then,***
   $23-x=\frac{1}{2}\cdot 11+12.$

If the middle number is odd, then I need to subtract two to get the odd integer before it, and I need to add two to get the odd integer after it.

6. The sum of three consecutive odd integers is 165. Let x be the middle of the three odd integers. Transcribe the statement accordingly.

   $(x-2)+x+(x+2)=165$

Write each of the following statements using symbolic language.

1. Bruce bought two books. One book costs $4.00 more than three times the other. Together, the two books cost him $72.

2. Janet is three years older than her sister Julie. Janet's brother is eight years younger than their sister Julie. The sum of all of their ages is 55 years.

3. The sum of three consecutive integers is 1,623.

4. One number is six more than another number. The sum of their squares is 90.

5. When you add 18 to $\frac{1}{4}$ of a number, you get the number itself.

6. When a fraction of 17 is taken away from 17, what remains exceeds one-third of seventeen by six.

7. The sum of two consecutive even integers divided by four is 189.5.

8. Subtract seven more than twice a number from the square of one-third of the number to get zero.

9. The sum of three consecutive integers is 42. Let x be the middle of the three integers. Transcribe the statement accordingly.

Exercises

Write each of the following statements in Exercises 1–12 as a mathematical expression. State whether or not the expression is linear or nonlinear. If it is nonlinear, then explain why.

1. The sum of a number and four times the number

2. The product of five and a number

3. Multiply six and the reciprocal of the quotient of a number and seven.

4. Twice a number subtracted from four times a number, added to 15

5. The square of the sum of six and a number

6. The cube of a positive number divided by the square of the same positive number

7. The sum of four consecutive numbers

8. Four subtracted from the reciprocal of a number

9. Half of the product of a number multiplied by itself three times

10. The sum that shows how many pages Maria read if she read 45 pages of a book yesterday and $\frac{2}{3}$ of the remaining pages today

11. An admission fee of $10 plus an additional $2 per game

12. Five more than four times a number and then twice that sum

Lesson Summary

A *linear expression* is an expression that is equivalent to the sum or difference of one or more expressions where each expression is either a number, a variable, or a product of a number and a variable.

A linear expression in x can be represented by terms whose variable x is raised to either a power of 0 or 1. For example, $4 + 3x$, $7x + x - 15$, and $\frac{1}{2}x + 7 - 2$ are all linear expressions in x. A nonlinear expression in x has terms where x is raised to a power that is not 0 or 1. For example, $2x^2 - 9$, $-6x^{-3} + 8 + x$, and $\frac{1}{x} + 8$ are all nonlinear expressions in x.

Name ____ Date ____________________

Write each of the following statements as a mathematical expression. State whether the expression is a linear or nonlinear expression in x.

1. Seven subtracted from five times a number, and then the difference added to nine times a number

2. Three times a number subtracted from the product of fifteen and the reciprocal of a number

3. Half of the sum of two and a number multiplied by itself three times

Write each of the following statements as a mathematical expression. State whether the expression is linear or nonlinear. If it is nonlinear, then explain why.

1. A number added to five cubed

   Let x be a number; then, $5^3 + x$ is a linear expression.

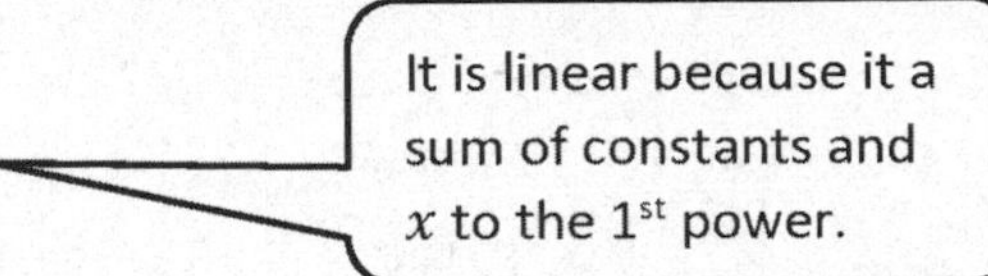

2. The quotient of seven and a number, added to twenty-five

   Let x be a number; then, $\frac{7}{x} + 25$ is a nonlinear expression.

   The term $\frac{7}{x}$ is the same as $7 \cdot \frac{1}{x}$ and $\frac{1}{x} = x^{-1}$, which is why it is not linear.

> I remember that $\frac{1}{x} = x^{-1}$ from the beginning of the year.

3. The sum that represents the number of hotdogs sold if 148 hotdogs were sold Thursday, half of the remaining hotdogs were sold on Friday, and 203 hotdogs were sold on Saturday

   Let x be the remaining number of hotdogs; then, $148 + \frac{1}{2}x + 203$ is a linear expression.

4. The product of 46 and a number, added to the reciprocal of the number squared

   Let x be a number; then, $46x + \frac{1}{x^2}$ is a nonlinear expression.

   The term $\frac{1}{x^2}$ is the same as x^{-2}, which is why it is not linear.

> I could write the expression as $\frac{1}{x^2} + 46x$ by applying the commutative property of addition.

5. The product of 12 and a number and then the product multiplied by itself seven times

   Let x be a number; then, $(12x)^7$ is a nonlinear expression. The expression can be written as $12^7 \cdot x^7$. The exponent of 7 with a base of x is the reason it is not linear.

6. The sum of seven and a number, multiplied by the number

   Let x be a number; then, $(7 + x)x$ is a nonlinear expression because $(7 + x)x = 7x + x^2$ after using the distributive property. It is nonlinear because the power of x in the term x^2 is greater than 1.

> I need to use parentheses around the sum of seven and a number.

Write each of the following statements as a mathematical expression. State whether the expression is linear or nonlinear. If it is nonlinear, then explain why.

1. A number decreased by three squared

2. The quotient of two and a number, subtracted from seventeen

3. The sum of thirteen and twice a number

4. 5.2 more than the product of seven and a number

5. The sum that represents the number of tickets sold if 35 tickets were sold Monday, half of the remaining tickets were sold on Tuesday, and 14 tickets were sold on Wednesday

6. The product of 19 and a number, subtracted from the reciprocal of the number cubed

7. The product of 15 and a number, and then the product multiplied by itself four times

8. A number increased by five and then divided by two

9. Eight times the result of subtracting three from a number

10. The sum of twice a number and four times a number subtracted from the number squared

11. One-third of the result of three times a number that is increased by 12

12. Five times the sum of one-half and a number

13. Three-fourths of a number multiplied by seven

14. The sum of a number and negative three, multiplied by the number

15. The square of the difference between a number and 10

Exercises

1. Is the equation a true statement when $x = -3$? In other words, is -3 a solution to the equation $6x + 5 = 5x + 8 + 2x$? Explain.

2. Does $x = 12$ satisfy the equation $16 - \frac{1}{2}x = \frac{3}{4}x + 1$? Explain.

3. Chad solved the equation $24x + 4 + 2x = 3(10x - 1)$ and is claiming that $x = 2$ makes the equation true. Is Chad correct? Explain.

4. Lisa solved the equation $x + 6 = 8 + 7x$ and claimed that the solution is $x = -\frac{1}{3}$. Is she correct? Explain.

5. Angel transformed the following equation from $6x + 4 - x = 2(x + 1)$ to $10 = 2(x + 1)$. He then stated that the solution to the equation is $x = 4$. Is he correct? Explain.

6. Claire was able to verify that $x = 3$ was a solution to her teacher's linear equation, but the equation got erased from the board. What might the equation have been? Identify as many equations as you can with a solution of $x = 3$.

7. Does an equation always have a solution? Could you come up with an equation that does not have a solution?

Lesson Summary

An equation is a statement about equality between two expressions. If the expression on the left side of the equal sign has the same value as the expression on the right side of the equal sign, then you have a true equation.

A solution of a linear equation in x is a number, such that when all instances of x are replaced with the number, the left side will equal the right side. For example, 2 is a solution to $3x + 4 = x + 8$ because when $x = 2$, the left side of the equation is

$$\begin{aligned} 3x + 4 &= 3(2) + 4 \\ &= 6 + 4 \\ &= 10, \end{aligned}$$

and the right side of the equation is

$$\begin{aligned} x + 8 &= 2 + 8 \\ &= 10. \end{aligned}$$

Since $10 = 10$, then $x = 2$ is a solution to the linear equation $3x + 4 = x + 8$.

Name ______________________________ Date ______________

1. Is 8 a solution to $\frac{1}{2}x + 9 = 13$? Explain.

2. Write three different equations that have $x = 5$ as a solution.

3. Is -3 a solution to the equation $3x - 5 = 4 + 2x$? Explain.

1. Given that $5x - 3 = 17$ and $7x + 3 = 17$, does $5x - 3 = 7x + 3$? Explain.

   yes, $5x - 3 = 7x + 3$ because a linear equation is a statement about equality. We are given that $5x - 3$ is equal to 17, but $7x + 3$ is also equal to 17. Since each linear expression is equal to the same number, the expressions are equal, $5x - 3 = 7x + 3$.

   > Since the left side of both expressions are equal to the same number, I can say that the expressions are equal to each other.

2. Is 5 a solution to the equation $3x - 1 = 5x + 7$? Explain.

   If we replace x with the number 5, then the left side of the equation is

   $$\begin{aligned} 3 \cdot (5) - 1 &= 15 - 1 \\ &= 14, \end{aligned}$$

   and the right side of the equation is

   $$\begin{aligned} 5 \cdot (5) + 7 &= 25 + 7 \\ &= 32. \end{aligned}$$

   Since $14 \neq 32$, 5 is not a solution of the equation $3x - 1 = 5x + 7$.

   > I need to see if the right side is equal to the left side when I replace x with the number 5. If the left side is not equal to the right side, then I know 5 is not a solution.

   > I know that a linear equation is really a question that is asking what number x will satisfy the equation.

3. Use the linear equation $11(x - 2) = 11x - 22$ to answer parts (a)–(c).

   a. Does $x = 3$ satisfy the equation above? Explain.

   if we replace x with the number 3, then the left side of the equation is

   $$\begin{aligned} 11(x - 2) &= 11(3 - 2) \\ &= 11(1) \\ &= 11, \end{aligned}$$

   and the right side of the equation is

   $$\begin{aligned} 11x - 22 &= 11 \cdot 3 - 22 \\ &= 33 - 22 \\ &= 11. \end{aligned}$$

   Since $11 = 11$, then $x = 3$ is a solution of the equation $11(x - 2) = 11x - 22$.

b. Is $x = -\frac{1}{2}$ a solution of the equation above? Explain.

If we replace x with the number $-\frac{1}{2}$, then the left side of the equation is

$$\begin{aligned} 11(x-2) &= 11\left(-\frac{1}{2}-2\right) \\ &= 11\left(-\frac{1}{2}-\frac{4}{2}\right) \\ &= 11\left(-\frac{5}{2}\right) \\ &= -\frac{55}{2}, \end{aligned}$$

I can rewrite 2 as an equivalent fraction with the same denominator as $\frac{1}{2}$.

and the right side of the equation is

$$\begin{aligned} 11x-22 &= 11\cdot -\frac{1}{2}-22 \\ &= -\frac{11}{2}-22 \\ &= -\frac{11}{2}-\frac{44}{2} \\ &= -\frac{55}{2}. \end{aligned}$$

Since the right side is equal to the left side, $-\frac{1}{2}$ is a solution.

Since $-\frac{55}{2} = -\frac{55}{2}$, $x = -\frac{1}{2}$ is a solution of the equation $11(x-2) = 11x - 22$.

c. What interesting fact about the equation $11(x-2) = 11x - 22$ is illuminated by the answers to parts (a) and (b)? Why do you think this is true?

I notice that the equation $11(x-2) = 11x - 22$ is an identity under the distributive law.

I remember my teacher saying illuminated means "What do I notice?"

I think I can choose any number for x and the equation will be true.

1. Given that $2x + 7 = 27$ and $3x + 1 = 28$, does $2x + 7 = 3x + 1$? Explain.

2. Is -5 a solution to the equation $6x + 5 = 5x + 8 + 2x$? Explain.

3. Does $x = 1.6$ satisfy the equation $6 - 4x = -\frac{x}{4}$? Explain.

4. Use the linear equation $3(x + 1) = 3x + 3$ to answer parts (a)–(d).

   a. Does $x = 5$ satisfy the equation above? Explain.

   b. Is $x = -8$ a solution of the equation above? Explain.

   c. Is $x = \frac{1}{2}$ a solution of the equation above? Explain.

   d. What interesting fact about the equation $3(x + 1) = 3x + 3$ is illuminated by the answers to parts (a), (b), and (c)? Why do you think this is true?

Exercises

For each problem, show your work, and check that your solution is correct.

1. Solve the linear equation $x + x + 2 + x + 4 + x + 6 = -28$. State the property that justifies your first step and why you chose it.

2. Solve the linear equation $2(3x + 2) = 2x - 1 + x$. State the property that justifies your first step and why you chose it.

3. Solve the linear equation $x - 9 = \frac{3}{5}x$. State the property that justifies your first step and why you chose it.

4. Solve the linear equation $29 - 3x = 5x + 5$. State the property that justifies your first step and why you chose it.

5. Solve the linear equation $\frac{1}{3}x - 5 + 171 = x$. State the property that justifies your first step and why you chose it.

Lesson Summary

The properties of equality, shown below, are used to transform equations into simpler forms. If A, B, C are rational numbers, then:

- If $A = B$, then $A + C = B + C$. Addition property of equality
- If $A = B$, then $A - C = B - C$. Subtraction property of equality
- If $A = B$, then $A \cdot C = B \cdot C$. Multiplication property of equality
- If $A = B$, then $\frac{A}{C} = \frac{B}{C}$, where C is not equal to zero. Division property of equality

To solve an equation, transform the equation until you get to the form of x equal to a constant ($x = 5$, for example).

Name ______________________________ Date ______________

1. Guess a number for x that would make the equation true. Check your solution.

$$5x - 2 = 8$$

2. Use the properties of equality to solve the equation $7x - 4 + x = 12$. State which property justifies your first step and why you chose it. Check your solution.

3. Use the properties of equality to solve the equation $3x + 2 - x = 11x + 9$. State which property justifies your first step and why you chose it. Check your solution.

For each problem, show your work, and check that your solution is correct.

1. Solve the linear equation $5x - 7 + 2x = -21$. State the property that justifies your first step and why you chose it.

   I used the commutative and distributive properties on the left side of the equal sign to simplify the expression to fewer terms.

$$5x - 7 + 2x = -21$$
$$5x + 2x - 7 = -21$$
$$(5 + 2)x - 7 = -21$$
$$7x - 7 = -21$$
$$7x - 7 + 7 = -21 + 7$$
$$7x = -14$$
$$\frac{1}{7}(7)x = \frac{1}{7}(-14)$$
$$x = -2$$

> The commutative property allows me to rearrange and group terms within expressions. The distributive property allows me to simplify expressions by combining terms that are alike.

Check: The left side is equal to $5(-2) - 7 + 2(-2) = -10 - 7 - 4 = -21$***, which is equal to the right side. Therefore,*** $x = -2$ ***is a solution to the equation*** $5x - 7 + 2x = -21$.

2. Solve the linear equation $\frac{1}{7}x - 11 = \frac{1}{4}x - 14$. State the property that justifies your first step and why you chose it.

I chose to use the addition property of equality to get all of the constants on one side of the equal sign and the subtraction property of equality to get all of the terms with an x on the other side of the equal sign.

$$\frac{1}{7}x - 11 = \frac{1}{4}x - 14$$
$$\frac{1}{7}x - 11 + 11 = \frac{1}{4}x - 14 + 11$$
$$\frac{1}{7}x - \frac{1}{4}x = \frac{1}{4}x - \frac{1}{4}x - 3$$
$$\left(\frac{1}{7} - \frac{1}{4}\right)x = -3$$
$$\left(\frac{4}{28} - \frac{7}{28}\right)x = -3$$
$$-\frac{3}{28}x = -3$$
$$-\frac{28}{3}\left(-\frac{3}{28}\right)x = -\frac{28}{3}(-3)$$
$$x = 28$$

I remember that the order doesn't matter, as long as I use the properties of equality correctly. I could use the subtraction property of equality to get all the terms with an x on one side of the equal sign and then use the addition property of equality to get all the constants on the other side.

Check: The left side of the equation is $\frac{1}{7}(28) - 11 = 4 - 11 = -7$. The right side of the equation is $\frac{1}{4}(28) - 14 = 7 - 14 = -7$. Since both sides equal -7, $x = 28$ is a solution to the equation $\frac{1}{7}x - 11 = \frac{1}{4}x - 14$.

I need to check my answer in the original equation because I may have made a mistake when transforming the equation.

3. Corey solved the linear equation $5x + 7 - 18x = 14 + 3x - 87$. His work is shown below. When he checked his answer, the left side of the equation did not equal the right side. Find and explain Corey's error, and then solve the equation correctly.

$$5x + 7 - 18x = 14 + 3x - 87$$
$$-13x + 7 = 3x - 73$$
$$-13x + 7 + 3x = 3x - 73 - 3x$$
$$-10x + 7 = -73$$
$$-10x + 7 - 7 = -73 - 7$$
$$-10x = -80$$
$$\frac{-10}{-10}x = \frac{-80}{-10}$$
$$x = 8$$

A strategy I used in class is to solve the linear equation and check my answer without looking at Corey's solution. I will compare my solution to Corey's to see if I find any differences.

Corey made a mistake on the third line. He added $3x$ to the left side of the equal sign and subtracted $3x$ on the right side of the equal sign. To use the property correctly, he should have subtracted $3x$ on both sides of the equal sign, making the equation at that point:

$$\mathbf{-13x + 7 - 3x = 3x - 73 - 3x}$$
$$\mathbf{-16x + 7 = -73}$$
$$\mathbf{-16x + 7 - 7 = -73 - 7}$$
$$\mathbf{-16x = -80}$$
$$\mathbf{\frac{-16}{-16}x = \frac{-80}{-16}}$$
$$\mathbf{x = 5}$$

For each problem, show your work, and check that your solution is correct.

1. Solve the linear equation $x + 4 + 3x = 72$. State the property that justifies your first step and why you chose it.

2. Solve the linear equation $x + 3 + x - 8 + x = 55$. State the property that justifies your first step and why you chose it.

3. Solve the linear equation $\frac{1}{2}x + 10 = \frac{1}{4}x + 54$. State the property that justifies your first step and why you chose it.

4. Solve the linear equation $\frac{1}{4}x + 18 = x$. State the property that justifies your first step and why you chose it.

5. Solve the linear equation $17 - x = \frac{1}{3} \cdot 15 + 6$. State the property that justifies your first step and why you chose it.

6. Solve the linear equation $\frac{x+x+2}{4} = 189.5$. State the property that justifies your first step and why you chose it.

7. Alysha solved the linear equation $2x - 3 - 8x = 14 + 2x - 1$. Her work is shown below. When she checked her answer, the left side of the equation did not equal the right side. Find and explain Alysha's error, and then solve the equation correctly.

$$
\begin{aligned}
2x - 3 - 8x &= 14 + 2x - 1 \\
-6x - 3 &= 13 + 2x \\
-6x - 3 + 3 &= 13 + 3 + 2x \\
-6x &= 16 + 2x \\
-6x + 2x &= 16 \\
-4x &= 16 \\
\frac{-4}{-4}x &= \frac{16}{-4} \\
x &= -4
\end{aligned}
$$

Example 1

One angle is five degrees less than three times the degree measure of another angle. Together, the angle measures have a sum of $143°$. What is the measure of each angle?

Example 2

Given a right triangle, find the degree measure of the angles if one angle is ten degrees more than four times the degree measure of the other angle and the third angle is the right angle.

Exercises

For each of the following problems, write an equation and solve.

1. A pair of congruent angles are described as follows: The degree measure of one angle is three more than twice a number, and the other angle's degree measure is 54.5 less than three times the number. Determine the measure of the angles in degrees.

2. The measure of one angle is described as twelve more than four times a number. Its supplement is twice as large. Find the measure of each angle in degrees.

3. A triangle has angles described as follows: The measure of the first angle is four more than seven times a number, the measure of the second angle is four less than the first, and the measure of the third angle is twice as large as the first. What is the measure of each angle in degrees?

4. One angle measures nine more than six times a number. A sequence of rigid motions maps the angle onto another angle that is described as being thirty less than nine times the number. What is the measure of the angle in degrees?

5. A right triangle is described as having an angle of measure six less than negative two times a number, another angle measure that is three less than negative one-fourth the number, and a right angle. What are the measures of the angles in degrees?

6. One angle is one less than six times the measure of another. The two angles are complementary angles. Find the measure of each angle in degrees.

Name ____ Date ____________________

For each of the following problems, write an equation and solve.

1. Given a right triangle, find the measures of all the angles, in degrees, if one angle is a right angle and the measure of the second angle is six less than seven times the measure of the third angle.

2. In a triangle, the measure of the first angle is six times a number. The measure of the second angle is nine less than the first angle. The measure of the third angle is three times the number more than the measure of the first angle. Determine the measure of each angle in degrees.

For each of the following problems, write an equation and solve.

The sum of the measures of complementary angles is $90°$.

1. An angle measures eleven more than four times a number. Its complement is two more than three times the number. What is the measure of each angle in degrees?

Let x be the number. Then, the measure of one angle is $4x + 11$. The measure of the other angle is $3x + 2$. Since the angles are complementary, the sum of their measures will be $90°$.

$$4x + 11 + 3x + 2 = 90$$
$$7x + 13 = 90$$
$$7x + 13 - 13 = 90 - 13$$
$$7x = 77$$
$$x = 11$$

I'm not done yet. I need to make sure I find the measure of each angle.

Replacing x with 11 in $4x + 11$ gives $4(11) + 11 = 44 + 11 = 55$.
Replacing x with 11 in $3x + 2$ gives $3(11) + 2 = 33 + 2 = 35$.
Therefore, the measures of the angles are $55°$ and $35°$.

2. The angles of a triangle are described as follows: $\angle A$ is the smallest angle. The measure of $\angle B$ is one more than the measure of $\angle A$ The measure of $\angle C$ is 3 more than twice the measure of $\angle A$. Find the measures of the three angles in degrees.

Let x be the measure of $\angle A$. Then, the measure of $\angle B$ is $x + 1°$ and $\angle C$ is $2x + 3°$. The sum of the measures of the angles must be $180°$.

$$x + x + 1° + 2x + 3° = 180°$$
$$4x + 4° = 180°$$
$$4x + 4° - 4° = 180° - 4°$$
$$4x = 176°$$
$$x = 44°$$

The sum of the measures of the interior angles of a triangle is $180°$.

The measures of the angles are as follows: $\angle A = 44°$, $\angle B = 45°$, and $\angle C = 2(44°) + 3° = 88° + 3° = 91°$.

3. A pair of corresponding angles are described as follows: The measure of one angle is fifteen less than four times a number, and the measure of the other angle is twenty more than four times the number. Are the angles congruent? Why or why not?

I need to use the fact that corresponding angles of parallel lines are congruent so that I can write an equation.

Let x be the number. Then, the measure of one angle is $4x - 15$, and the measure of the other angle is $4x + 20$. Assume they are congruent, which means their measures are equal.

$$4x - 15 = 4x + 20$$
$$4x - 4x - 15 = 4x - 4x + 20$$
$$-15 \neq 20$$

Since $-15 \neq 20$, the angles are not congruent.

4. Three angles are described as follows: $\angle A$ is one-third the size of $\angle B$. The measure of $\angle C$ is equal to seven more than three times the measure of $\angle B$. The sum of the measures of $\angle A$ and $\angle C$ is $147°$. Can the three angles form a triangle? Why or why not?

Since I don't know if the three angle measures form a triangle, I need to use the sum of the two triangles to write my equation.

Let x represent the measure of $\angle B$. Then, the measure of $\angle A$ is $\frac{x}{3}$ and the measure of $\angle C$ is $3x + 7°$.

The sum of the measures of $\angle A$ and $\angle C$ is $147°$.

$$\frac{x}{3} + 3x + 7° = 147°$$
$$\frac{1}{3}x + \frac{9}{3}x + 7° = 147°$$
$$\left(\frac{1}{3} + \frac{9}{3}\right)x + 7° = 147°$$
$$\frac{10}{3}x + 7° - 7° = 147° - 7°$$
$$\frac{10}{3}x = 140°$$
$$10x = 420°$$
$$x = 42°$$

I need to check the sum of the three angles to see if they form a triangle.

The measure of $\angle A$ is $\left(\frac{42}{3}\right)^{\circ} = 14°$, the measure of $\angle B$ is $42°$, and the measure of $\angle C$ is $3(42°) + 7° = 133°$. The sum of the three angles is $14° + 42° + 133° = 189°$. Since the sum of the measures of the interior angles of a triangle must have a sum of $180°$, these angles do not form a triangle. Their sum is too large.

For each of the following problems, write an equation and solve.

1. The measure of one angle is thirteen less than five times the measure of another angle. The sum of the measures of the two angles is $140°$. Determine the measure of each angle in degrees.

2. An angle measures seventeen more than three times a number. Its supplement is three more than seven times the number. What is the measure of each angle in degrees?

3. The angles of a triangle are described as follows: $\angle A$ is the largest angle; its measure is twice the measure of $\angle B$. The measure of $\angle C$ is 2 less than half the measure of $\angle B$. Find the measures of the three angles in degrees.

4. A pair of corresponding angles are described as follows: The measure of one angle is five less than seven times a number, and the measure of the other angle is eight more than seven times the number. Are the angles congruent? Why or why not?

5. The measure of one angle is eleven more than four times a number. Another angle is twice the first angle's measure. The sum of the measures of the angles is $195°$. What is the measure of each angle in degrees?

6. Three angles are described as follows: $\angle B$ is half the size of $\angle A$. The measure of $\angle C$ is equal to one less than two times the measure of $\angle B$. The sum of $\angle A$ and $\angle B$ is $114°$. Can the three angles form a triangle? Why or why not?

Exercises

Find the value of x that makes the equation true.

1. $17 - 5(2x - 9) = -(-6x + 10) + 4$

2. $-(x - 7) + \frac{5}{3} = 2(x + 9)$

3. $\frac{4}{9} + 4(x - 1) = \frac{28}{9} - (x - 7x) + 1$

4. $5(3x + 4) - 2x = 7x - 3(-2x + 11)$

5. $7x - (3x + 5) - 8 = \frac{1}{2}(8x + 20) - 7x + 5$

6. Write at least three equations that have no solution.

Lesson Summary

The distributive property is used to expand expressions. For example, the expression $2(3x - 10)$ is rewritten as $6x - 20$ after the distributive property is applied.

The distributive property is used to simplify expressions. For example, the expression $7x + 11x$ is rewritten as $(7 + 11)x$ and $18x$ after the distributive property is applied.

The distributive property is applied only to terms within a group:

$$4(3x + 5) - 2 = 12x + 20 - 2.$$

Notice that the term -2 is not part of the group and, therefore, not multiplied by 4.

When an equation is transformed into an untrue sentence, such as $5 \neq 11$, we say the equation has no solution.

Name ______________________________ Date ______________

Transform the equation if necessary, and then solve to find the value of x that makes the equation true.

1. $5x - (x + 3) = \frac{1}{3}(9x + 18) - 5$

2. $5(3x + 9) - 2x = 15x - 2(x - 5)$

1. $3x - (x + 2) + 11x = \frac{1}{2}(4x - 8)$

The negative sign in front of the parentheses means to take the opposite of each term inside the parentheses.

$$\begin{aligned} 3x - (x + 2) + 11x &= \frac{1}{2}(4x - 8) \\ 3x - x - 2 + 11x &= 2x - 4 \\ 13x - 2 &= 2x - 4 \\ 13x - 2x - 2 &= 2x - 2x - 4 \\ 11x - 2 &= -4 \\ 11x - 2 + 2 &= -4 + 2 \\ 11x &= -2 \\ \frac{11}{11}x &= -\frac{2}{11} \\ x &= -\frac{2}{11} \end{aligned}$$

I need to use the distributive property to each term inside the parentheses. It will allow me to see all the terms and collect like terms.

I need to check my answer.

Check: The left side is $3\left(-\frac{2}{11}\right) - \left(-\frac{2}{11} + 2\right) + 11\left(-\frac{2}{11}\right) = -\frac{6}{11} - \frac{20}{11} - 2 = -\frac{48}{11}$.

The right side is $\frac{1}{2}\left(4\left(-\frac{2}{11}\right) - 8\right) = \frac{1}{2}\left(-\frac{8}{11} - \frac{88}{11}\right) = \frac{1}{2}\left(\frac{96}{11}\right) = -\frac{48}{11}$. ***Since*** $-\frac{48}{11} = -\frac{48}{11}$,

$x = -\frac{2}{11}$ ***is the solution.***

2. $5(2 + x) - 4 = 81$

$$\begin{aligned} 5(2 + x) - 4 &= 81 \\ 10 + 5x - 4 &= 81 \\ 5x + 6 &= 81 \\ 5x + 6 - 6 &= 81 - 6 \\ 5x &= 75 \\ x &= 15 \end{aligned}$$

I need to use the distributive property to each term inside the parentheses only but not to the -4.

I can check this answer mentally.

3. $6x + \frac{1}{3}(9x + 5) = 10x + \frac{13}{3} - (x + 1)$

$$\mathbf{6x + \frac{1}{3}(9x + 5) = 10x + \frac{13}{3} - (x + 1)}$$
$$\mathbf{6x + 3x + \frac{5}{3} = 10x + \frac{13}{3} - x - 1}$$
$$\mathbf{9x + \frac{5}{3} = 9x + \frac{10}{3}}$$
$$\mathbf{9x - 9x + \frac{5}{3} = 9x - 9x + \frac{10}{3}}$$
$$\mathbf{\frac{5}{3} \neq \frac{10}{3}}$$

This is an untrue sentence; therefore, this equation has no solution.

This equation has no solution.

Transform the equation if necessary, and then solve it to find the value of x that makes the equation true.

1. $x - (9x - 10) + 11 = 12x + 3\left(-2x + \frac{1}{3}\right)$

2. $7x + 8\left(x + \frac{1}{4}\right) = 3(6x - 9) - 8$

3. $-4x - 2(8x + 1) = -(-2x - 10)$

4. $11(x + 10) = 132$

5. $37x + \frac{1}{2} - \left(x + \frac{1}{4}\right) = 9(4x - 7) + 5$

6. $3(2x - 14) + x = 15 - (-9x - 5)$

7. $8(2x + 9) = 56$

Exercises

Solve each of the following equations for x.

1. $7x - 3 = 5x + 5$

2. $7x - 3 = 7x + 5$

3. $7x - 3 = -3 + 7x$

Give a brief explanation as to what kind of solution(s) you expect the following linear equations to have. Transform the equations into a simpler form if necessary.

4. $11x - 2x + 15 = 8 + 7 + 9x$

5. $3(x - 14) + 1 = -4x + 5$

6. $-3x + 32 - 7x = -2(5x + 10)$

7. $\frac{1}{2}(8x + 26) = 13 + 4x$

8. Write two equations that have no solutions.

9. Write two equations that have one unique solution each.

10. Write two equations that have infinitely many solutions.

Lesson Summary

There are three classifications of solutions to linear equations: one solution (unique solution), no solution, or infinitely many solutions.

Equations with no solution will, after being simplified, have coefficients of x that are the same on both sides of the equal sign and constants that are different. For example, $x + b = x + c$, where b and c are constants that are not equal. A numeric example is $8x + 5 = 8x - 3$.

Equations with infinitely many solutions will, after being simplified, have coefficients of x and constants that are the same on both sides of the equal sign. For example, $x + a = x + a$, where a is a constant. A numeric example is $6x + 1 = 1 + 6x$.

Name ____________________ Date ____________

Give a brief explanation as to what kind of solution(s) you expect the following linear equations to have. Transform the equations into a simpler form if necessary.

1. $3(6x + 8) = 24 + 18x$

2. $12(x + 8) = 11x - 5$

3. $5x - 8 = 11 - 7x + 12x$

1. Give a brief explanation as to what kind of solution(s) you expect for the linear equation $12x + 7 = -3(9 - 5x)$. Transform the equation into a simpler form if necessary.

The coefficients of x are different and so are the constants.

$$12x + 7 = -3(9 - 5x)$$
$$12x + 7 = -27 + 15x$$

This equation will have a unique solution.

After I use the distributive property on the right side, the coefficients of x are different $(12 \neq 15)$, and the constants are different $(7 \neq -27)$ on each side. This means the equation will have a unique solution.

2. Give a brief explanation as to what kind of solution(s) you expect for the linear equation $18\left(\frac{1}{2} + \frac{1}{3}x\right) = 6x + 9$. Transform the equation into a simpler form if necessary.

$$18\left(\frac{1}{2} + \frac{1}{3}x\right) = 6x + 9$$
$$9 + 6x = 6x + 9$$

After I use the distributive property on the left side, the coefficients of x are the same $(6 = 6)$, and the constants are the same $(9 = 9)$ on each side. This means the equation will have infinitely many solutions.

This is an identity under the distributive property. Therefore, this equation will have infinitely many solutions.

3. Give a brief explanation as to what kind of solution(s) you expect for the linear equation $5(2x + 4) = 2(5x - 10)$. Transform the equation into a simpler form if necessary.

$$5(2x + 4) = 2(5x - 10)$$
$$10x + 20 = 10x - 20$$

After I use the distributive property on both sides of the equation, the coefficients of x are the same $(10 = 10)$, and the constants are different $(20 \neq -20)$ on each side. This means the equation will have no solution.

The coefficients of x are the same, but the constants are different. Therefore, this equation has no solutions.

1. Give a brief explanation as to what kind of solution(s) you expect for the linear equation $18x + \frac{1}{2} = 6(3x + 25)$. Transform the equation into a simpler form if necessary.

2. Give a brief explanation as to what kind of solution(s) you expect for the linear equation $8 - 9x = 15x + 7 + 3x$. Transform the equation into a simpler form if necessary.

3. Give a brief explanation as to what kind of solution(s) you expect for the linear equation $5(x + 9) = 5x + 45$. Transform the equation into a simpler form if necessary.

4. Give three examples of equations where the solution will be unique; that is, only one solution is possible.

5. Solve one of the equations you wrote in Problem 4, and explain why it is the only solution.

6. Give three examples of equations where there will be no solution.

7. Attempt to solve one of the equations you wrote in Problem 6, and explain why it has no solution.

8. Give three examples of equations where there will be infinitely many solutions.

9. Attempt to solve one of the equations you wrote in Problem 8, and explain why it has infinitely many solutions.

Example 3

Can this equation be solved?

$$\frac{6+x}{7x+\frac{2}{3}}=\frac{3}{8}$$

Example 4

Can this equation be solved?

$$\frac{7}{3x+9}=\frac{1}{8}$$

Example 5

In the diagram below, $\Delta\ ABC \sim \Delta\ A'B'C'$. Using what we know about similar triangles, we can determine the value of x.

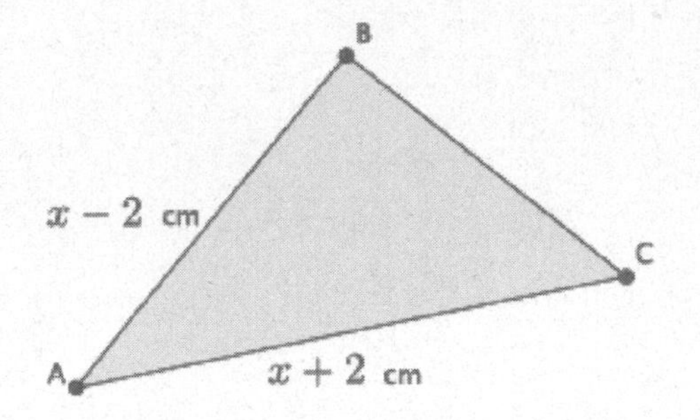

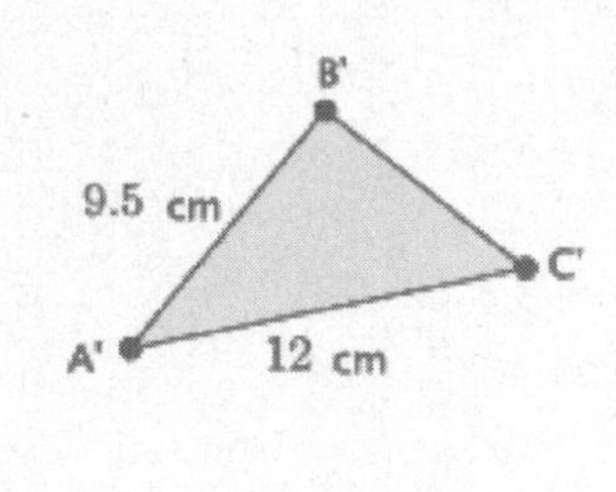

Exercises

Solve the following equations of rational expressions, if possible.

1. $\dfrac{2x+1}{9} = \dfrac{1-x}{6}$

2. $\dfrac{5+2x}{3x-1}=\dfrac{6}{7}$

3. $\dfrac{x+9}{12}=\dfrac{-2x-\frac{1}{2}}{3}$

4. $\dfrac{8}{3-4x}=\dfrac{5}{2x+\frac{1}{4}}$

Lesson Summary

Some proportions are linear equations in disguise and are solved the same way we normally solve proportions.

When multiplying a fraction with more than one term in the numerator and/or denominator by a number, put the expressions with more than one term in parentheses so that you remember to use the distributive property when transforming the equation. For example:

$$\frac{x+4}{2x-5} = \frac{3}{5}$$
$$5(x+4) = 3(2x-5).$$

The equation $5(x+4) = 3(2x-5)$ is now clearly a linear equation and can be solved using the properties of equality.

Name ____________________________ Date ____________

Solve the following equations for x.

1. $\frac{5x-8}{3}=\frac{11x-9}{5}$

2. $\frac{x+11}{7}=\frac{2x+1}{-8}$

3. $\frac{-x-2}{-4}=\frac{3x+6}{2}$

Solve the following equations of rational expressions, if possible. If the equation cannot be solved, explain why.

1. $\frac{x+5}{-2} = \frac{3-x}{7}$

$$\frac{x+5}{-2} = \frac{3-x}{7}$$
$$-2(3-x) = (x+5)7$$
$$-6+2x = 7x+35$$
$$-6+2x-2x = 7x-2x+35$$
$$-6 = 5x+35$$
$$-6-35 = 5x+35-35$$
$$-41 = 5x$$
$$-\frac{41}{5} = x$$

I can multiply each numerator by the other fraction's denominator. I put the expressions with more than one term in parentheses so that I remember to use the distributive property.

2. $\frac{12}{x-3} = \frac{4}{x+2}$

$$\frac{12}{x-3} = \frac{4}{x+2}$$
$$12(x+2) = (x-3)4$$
$$12x+24 = 4x-12$$
$$12x-4x+24 = 4x-4x-12$$
$$8x+24 = -12$$
$$8x+24-24 = -12-24$$
$$8x = -36$$
$$\frac{8}{8}x = -\frac{36}{8}$$
$$x = -\frac{9}{2}$$

I used the distributive property, and now I see that I have a linear equation that I can solve using properties of equalities that I learned earlier in the module in Lesson 4.

I can rewrite the fraction $-\frac{36}{8}$ as $-\frac{9}{2}$ because they are equivalent.

3. $\dfrac{\frac{1}{3}x-2}{8}=\dfrac{4x}{9}$

$$\frac{\frac{1}{3}x-2}{8}=\frac{4x}{9}$$
$$\left(\frac{1}{3}x-2\right)9=8(4x)$$
$$3x-18=32x$$
$$3x-3x-18=32x-3x$$
$$-18=29x$$
$$-\frac{18}{29}=x$$

> I could write the equation as $8(4x)=9\left(\frac{1}{3}x-2\right)$ because when I distribute, I will get $32x=3x-18$. When I use the properties of equalities, my answer will be the same, $x=-\frac{18}{29}$.

4. In the diagram below, $\Delta\ ABC \sim \Delta\ A'\ B'\ C'$. Determine the lengths of $\overline{AB}$ and $\overline{AC}$.

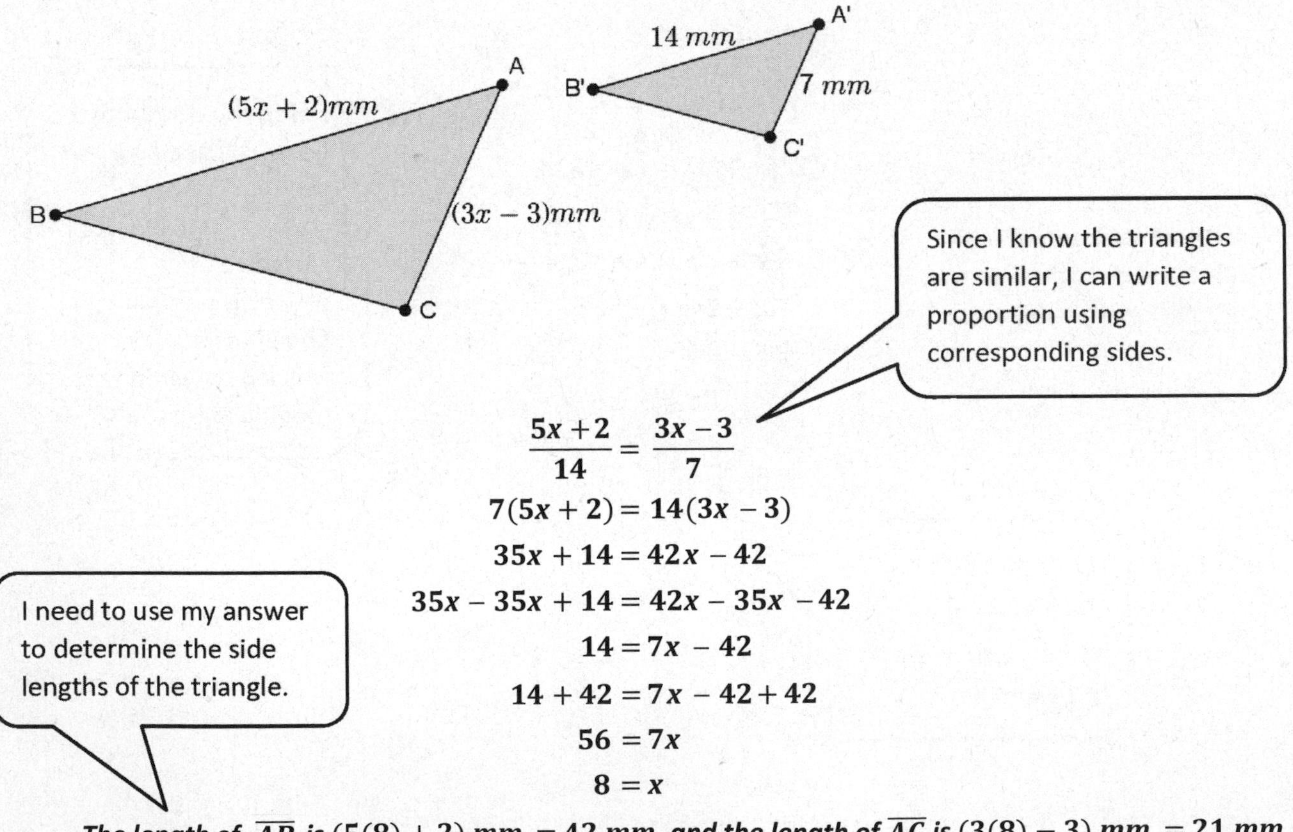

> Since I know the triangles are similar, I can write a proportion using corresponding sides.

$$\frac{5x+2}{14}=\frac{3x-3}{7}$$
$$7(5x+2)=14(3x-3)$$
$$35x+14=42x-42$$
$$35x-35x+14=42x-35x-42$$
$$14=7x-42$$
$$14+42=7x-42+42$$
$$56=7x$$
$$8=x$$

> I need to use my answer to determine the side lengths of the triangle.

The length of $\overline{AB}$ is $(5(8)+2)$ mm $=42$ mm, and the length of $\overline{AC}$ is $(3(8)-3)$ mm $=21$ mm.

Solve the following equations of rational expressions, if possible. If an equation cannot be solved, explain why.

1. $\frac{5}{6x-2} = \frac{-1}{x+1}$

2. $\frac{4-x}{8} = \frac{7x-1}{3}$

3. $\frac{3x}{x+2} = \frac{5}{9}$

4. $\frac{\frac{1}{2}x+6}{3} = \frac{x-3}{2}$

5. $\frac{7-2x}{6} = \frac{x-5}{1}$

6. $\frac{2x+5}{2} = \frac{3x-2}{6}$

7. $\frac{6x+1}{3} = \frac{9-x}{7}$

8. $\frac{\frac{1}{3}x-8}{12} = \frac{-2-x}{15}$

9. $\frac{3-x}{1-x} = \frac{3}{2}$

10. In the diagram below, $\Delta\, ABC \sim \Delta\, A'B'\,C'$. Determine the lengths of $\overline{AC}$ and $\overline{BC}$.

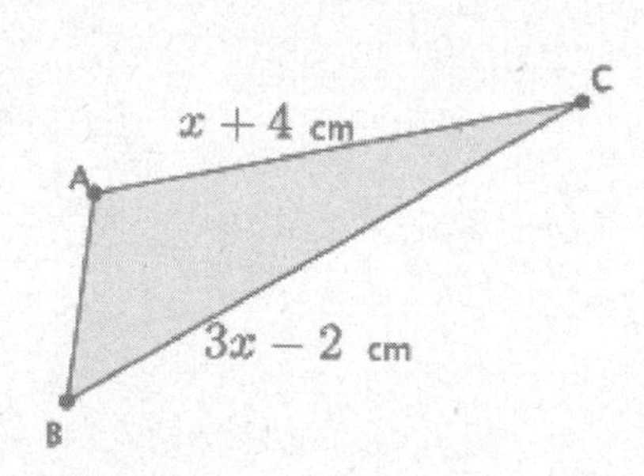

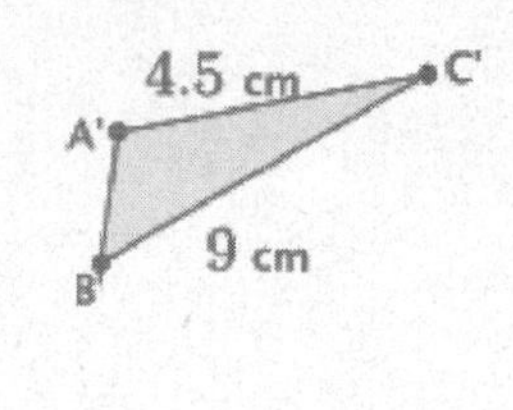

Exercises

1. Write the equation for the 15th step.

2. How many people would see the photo after 15 steps? Use a calculator if needed.

3. Marvin paid an entrance fee of $\$5$ plus an additional $\$1.25$ per game at a local arcade. Altogether, he spent $\$26.25$. Write and solve an equation to determine how many games Marvin played.

4. The sum of four consecutive integers is -26. What are the integers?

5. A book has x pages. How many pages are in the book if Maria read 45 pages of a book on Monday, $\frac{1}{2}$ the book on Tuesday, and the remaining 72 pages on Wednesday?

6. A number increased by 5 and divided by 2 is equal to 75. What is the number?

7. The sum of thirteen and twice a number is seven less than six times a number. What is the number?

8. The width of a rectangle is 7 less than twice the length. If the perimeter of the rectangle is 43.6 inches, what is the area of the rectangle?

9. Two hundred fifty tickets for the school dance were sold. On Monday, 35 tickets were sold. An equal number of tickets were sold each day for the next five days. How many tickets were sold on one of those days?

10. Shonna skateboarded for some number of minutes on Monday. On Tuesday, she skateboarded for twice as many minutes as she did on Monday, and on Wednesday, she skateboarded for half the sum of minutes from Monday and Tuesday. Altogether, she skateboarded for a total of three hours. How many minutes did she skateboard each day?

11. In the diagram below, $\Delta ABC \sim \Delta A'B'C'$. Determine the length of $\overline{AC}$ and $\overline{BC}$.

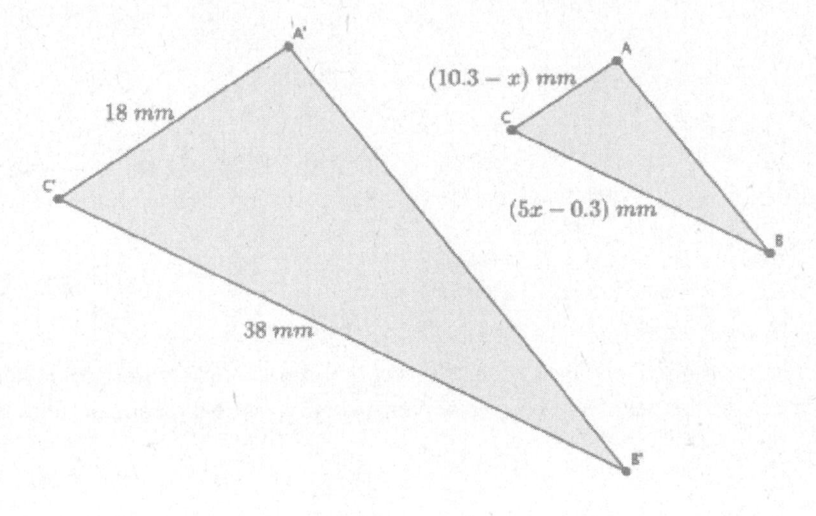

Name ______________________________ Date ______________

1. Rewrite the equation that would represent the sum in the fifth step of the Facebook problem:

$$S_5 = 7 + 7 \cdot 5 + 7 \cdot 5^2 + 7 \cdot 5^3 + 7 \cdot 5^4.$$

2. The sum of four consecutive integers is 74. Write an equation, and solve to find the numbers.

1. You forward a blog that you found online to five of your friends. They liked it so much that they each forwarded it on to two of their friends, who then each forwarded it on to two of their friends, and so on. The number of people who saw the blog is shown below. Let S_1 represent the number of people who saw the blog after one step, let S_2 represent the number of people who saw the blog after two steps, and so on.

$$S_1 = 5$$
$$S_2 = 5 + 5 \cdot 2$$
$$S_3 = 5 + 5 \cdot 2 + 5 \cdot 2^2$$
$$S_4 = 5 + 5 \cdot 2 + 5 \cdot 2^2 + 5 \cdot 2^3$$

I will start with S_2 since $S_1 = 5$ and try to manipulate S_2 into an equation that contains S_2.

a. Find the pattern in the equations.

By adding $5 \cdot 2^2$, I can use the distributive property to get a linear equation in S_2.

$$S_2 = 5 + 5 \cdot 2$$
$$S_2 - 5 = 5 \cdot 2$$
$$S_2 - 5 + 5 \cdot 2^2 = 5 \cdot 2 + 5 \cdot 2^2$$
$$S_2 - 5 + 5 \cdot 2^2 = 2(5 + 5 \cdot 2)$$
$$S_2 - 5 + 5 \cdot 2^2 = 2S_2$$

$$S_3 = 5 + 5 \cdot 2 + 5 \cdot 2^2$$
$$S_3 - 5 = 5 \cdot 2 + 5 \cdot 2^2$$
$$S_3 - 5 + 5 \cdot 2^3 = 5 \cdot 2 + 5 \cdot 2^2 + 5 \cdot 2^3$$
$$S_3 - 5 + 5 \cdot 2^3 = 2(5 + 5 \cdot 2 + 5 \cdot 2^2)$$
$$S_3 - 5 + 5 \cdot 2^3 = 2S_3$$

By adding $5 \cdot 2$ raised to the power of the step number, I can use the distributive property to get a linear equation in terms of that step number.

I don't want to multiply out any of the terms so that I can see the pattern better.

$$S_4 = 5 + 5 \cdot 2 + 5 \cdot 2^2 + 5 \cdot 2^3$$
$$S_4 - 5 = 5 \cdot 2 + 5 \cdot 2^2 + 5 \cdot 2^3$$
$$S_4 - 5 + 5 \cdot 2^4 = 5 \cdot 2 + 5 \cdot 2^2 + 5 \cdot 2^3 + 5 \cdot 2^4$$
$$S_4 - 5 + 5 \cdot 2^4 = 2(5 + 5 \cdot 2 + 5 \cdot 2^2 + 5 \cdot 2^3)$$
$$S_4 - 5 + 5 \cdot 2^4 = 2S_4$$

b. Assuming the trend continues, how many people will have seen the blog after 8 steps?

I want to use the properties of equality to get S_8 on one side and the constants on the other side of the equal sign and use the distributive property.

$$S_8 - 5 + 5 \cdot 2^8 = 2S_8$$
$$S_8 - 2S_8 = 5 - 5 \cdot 2^8$$
$$S_8(1-2) = 5 - 5 \cdot 2^8$$
$$S_8(1-2) = 5(1-2^8)$$
$$S_8 = \frac{5(1-2^8)}{(1-2)}$$

I multiplied out on the last step.

$$S_8 = 1,275$$

After** 8 **steps,** $1{,}275$ **people will have seen the blog.

c. How many people will have seen the blog after n steps?

$$S_n = \frac{5(1-2^n)}{(1-2)}$$

I see a pattern from the work I have done.

2. The length of a rectangle is 4 more than 2 times the width. If the perimeter of the rectangle is 20.6 cm. what is the area of the rectangle?

Since I know the perimeter, I will write my equation in terms of perimeter. Perimeter of a rectangle means I need to add twice the width to twice the length, $P = 2w + 2l$.

***Let x represent the width of the rectangle. Then the length of the rectangle is** $4 + 2x$.*

$$2(4 + 2x) + 2x = 20.6$$
$$8 + 4x + 2x = 20.6$$
$$8 + 6x = 20.6$$
$$6x = 12.6$$
$$x = \frac{12.6}{6}$$
$$x = 2.1$$

The problem asked for the area of the rectangle. Area of a rectangle means I have to multiply the length and width.

***The width of the rectangle is** 2.1 cm**, and the length is** $(4 + 2(2.1))$ cm $=$ 8.2 cm**, so the area is** 17.22 cm^2.*

3. Each month, Gilbert pays \$42 to his phone company just to use the phone. Each text he sends costs him an additional \$0.15. In June, his phone bill was \$162.75. In July, his phone bill was \$155.85. How many texts did he send each month?

Let x be the number of texts he sent in June.

$$42 + 0.15x = 162.75$$
$$0.15x = 120.75$$
$$x = \frac{120.75}{0.15}$$
$$x = 805$$

He sent 805 texts in June.

Let y be the number of texts he sent in July.

$$42 + 0.15y = 155.85$$
$$0.15y = 113.85$$
$$y = \frac{113.85}{0.15}$$
$$y = 759$$

He sent 759 texts in July.

I am using a different letter to define my variable because the number of texts for July is different than the number of texts for June since the cost was different for both months.

4. In the diagram below, $\Delta\ ABC \sim \Delta\ A'B'\ C'$. Determine the measure of $\angle A$.

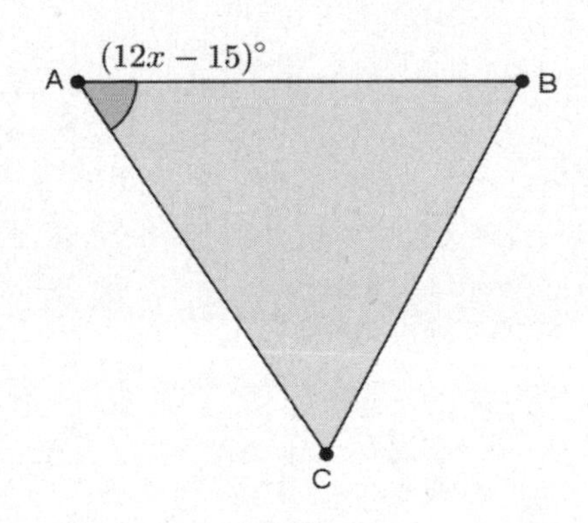

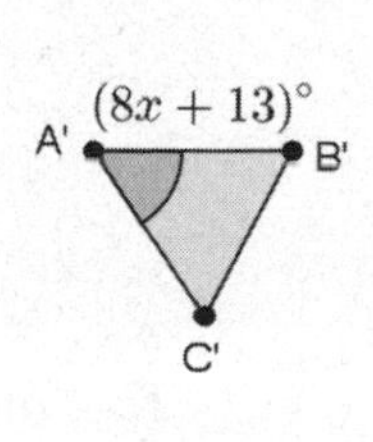

Since the triangles are similar, the angles are equal in measure.

$$12x - 15 = 8x + 13$$
$$12x - 8x - 15 = 8x - 8x + 13$$
$$4x - 15 = 13$$
$$4x - 15 + 15 = 13 + 15$$
$$4x = 28$$
$$x = 7$$

The measure of $\angle A$ is $(12(7) - 15)^\circ = 69^\circ$.

1. You forward an e-card that you found online to three of your friends. They liked it so much that they forwarded it on to four of their friends, who then forwarded it on to four of their friends, and so on. The number of people who saw the e-card is shown below. Let S_1 represent the number of people who saw the e-card after one step, let S_2 represent the number of people who saw the e-card after two steps, and so on.

$$S_1 = 3$$
$$S_2 = 3 + 3 \cdot 4$$
$$S_3 = 3 + 3 \cdot 4 + 3 \cdot 4^2$$
$$S_4 = 3 + 3 \cdot 4 + 3 \cdot 4^2 + 3 \cdot 4^3$$

   a. Find the pattern in the equations.

   b. Assuming the trend continues, how many people will have seen the e-card after 10 steps?

   c. How many people will have seen the e-card after n steps?

For each of the following questions, write an equation, and solve to find each answer.

2. Lisa has a certain amount of money. She spent $\$39$ and has $\frac{3}{4}$ of the original amount left. How much money did she have originally?

3. The length of a rectangle is 4 more than 3 times the width. If the perimeter of the rectangle is 18.4 cm, what is the area of the rectangle?

4. Eight times the result of subtracting 3 from a number is equal to the number increased by 25. What is the number?

5. Three consecutive odd integers have a sum of 3. What are the numbers?

6. Each month, Liz pays $\$35$ to her phone company just to use the phone. Each text she sends costs her an additional $\$0.05$. In March, her phone bill was $\$72.60$. In April, her phone bill was $\$65.85$. How many texts did she send each month?

7. Claudia is reading a book that has 360 pages. She read some of the book last week. She plans to read 46 pages today. When she does, she will be $\frac{4}{5}$ of the way through the book. How many pages did she read last week?

8. In the diagram below, $\Delta ABC \sim \Delta A'B'C'$. Determine the measure of $\angle A$.

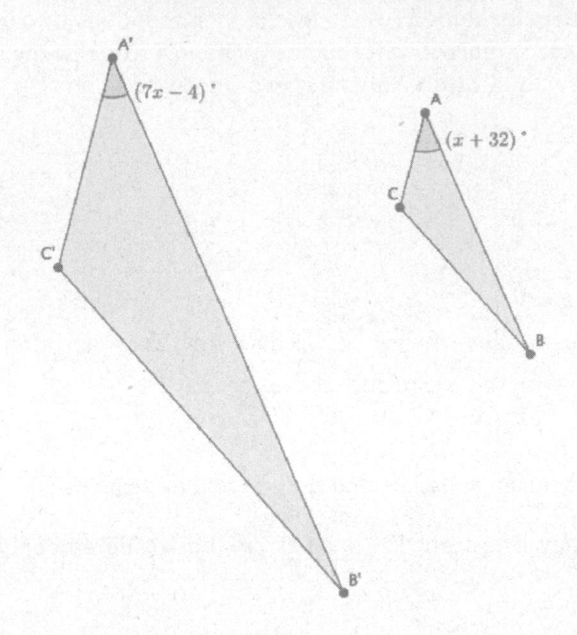

9. In the diagram below, $\Delta ABC \sim \Delta A'B'C'$. Determine the measure of $\angle A$.

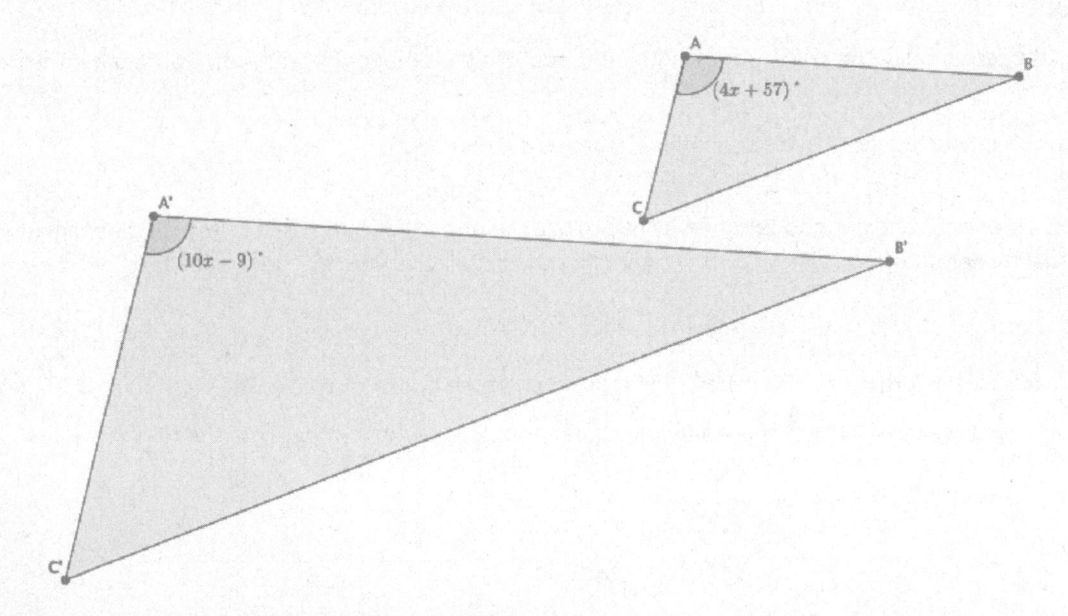

Example 1

Paul walks 2 miles in 25 minutes. How many miles can Paul walk in 137.5 minutes?

Time (in minutes)	Distance (in miles)
25	2

Exercises

1. Wesley walks at a constant speed from his house to school 1.5 miles away. It took him 25 minutes to get to school.
   a. What fraction represents his constant speed, C?

   b. You want to know how many miles he has walked after 15 minutes. Let y represent the distance he traveled after 15 minutes of walking at the given constant speed. Write a fraction that represents the constant speed, C, in terms of y.

   c. Write the fractions from parts (a) and (b) as a proportion, and solve to find how many miles Wesley walked after 15 minutes.

   d. Let y be the distance in miles that Wesley traveled after x minutes. Write a linear equation in two variables that represents how many miles Wesley walked after x minutes.

2. Stefanie drove at a constant speed from her apartment to her friend's house 20 miles away. It took her 45 minutes to reach her destination.
   a. What fraction represents her constant speed, C?

b. What fraction represents constant speed, C, if it takes her x number of minutes to get halfway to her friend's house?

c. Write and solve a proportion using the fractions from parts (a) and (b) to determine how many minutes it takes her to get to the halfway point.

d. Write a two-variable equation to represent how many miles Stefanie can drive over any time interval.

3. The equation that represents how many miles, y, Dave is from Town A for any given time in hours x is $y = 50x + 15$. Use the equation to complete the table below.

x (hours)	Linear Equation: $y = 50x + 15$	y (miles)
1	$y = 50(1) + 15$	65
2		
3		
3.5		
4.1		

Lesson Summary

Average speed is found by taking the total distance traveled in a given time interval, divided by the time interval.

If y is the total distance traveled in a given time interval x, then $\frac{y}{x}$ is the average speed.

If we assume the same average speed over any time interval, then we have constant speed, which can then be used to express a linear equation in two variables relating distance and time.

If $\frac{y}{x} = C$, where C is a constant, then you have constant speed.

Name ____ Date ____________________

Alex skateboards at a constant speed from his house to school 3.8 miles away. It takes him 18 minutes.

a. What fraction represents his constant speed, C?

b. After school, Alex skateboards at the same constant speed to his friend's house. It takes him 10 minutes. Write the fraction that represents constant speed, C, if he travels a distance of y.

c. Write the fractions from parts (a) and (b) as a proportion, and solve to find out how many miles Alex's friend's house is from school. Round your answer to the tenths place.

1. Jurgen types a paper for his Humanities class at a constant speed. He types 12 pages, and it took him 66 minutes.

   a. What fraction represents his constant speed, C?

   To write the fraction for his constant speed, I have to compare the number of pages typed to the interval of time spent typing.

   $$C = \frac{12}{66} = \frac{2}{11}$$

   b. Write the fraction that represents his constant speed, C, if he types y pages in 24 minutes.

   $$C = \frac{y}{24}$$

   c. Write a proportion using the fractions from parts (a) and (b) to determine how many pages he types after 24 minutes. Round your answer to the hundredths place.

   $$\frac{2}{11} = \frac{y}{24}$$
   $$2(24) = 11(y)$$
   $$48 = 11(y)$$
   $$\frac{1}{11}(48) = \frac{1}{11}(11)y$$
   $$4.36 \approx y$$

   Jurgen types approximately 4.36 pages in 24 minutes.

   d. Write a two-variable equation to represent how many pages Jurgen can type over any time interval.

   Let y represent the number of pages typed. Let x represent the number of minutes typed.

   When I write a two-variable equation, I have to remember to define my variables.

   $$\frac{2}{11} = \frac{y}{x}$$
   $$2(x) = 11(y)$$
   $$\frac{1}{11}(2)x = \frac{1}{11}(11)y$$
   $$\frac{2}{11}x = y$$

2. Parker runs at a constant speed of 6.25 miles per hour.

   a. If he runs for y miles and it takes him x hours, write the two-variable equation to represent the number of miles Parker can run in x hours.

   Let y represent the number of miles run. Let x represent the number of hours run.

$$\frac{6.25}{1} = \frac{y}{x}$$
$$6.25x = y$$

   b. Parker has been training for a marathon by running to the school 11 miles from his house, then to the park 2 miles from the school, and then returning home, which is 14 miles from the park. Assuming he runs at a constant speed the entire time, how long will it take him to get back home after running his route? Round your answer to the hundredths place.

   Total miles: $11 + 2 + 14 = 27$. Let x be the number of hours run.

$$6.25x = 27$$
$$\frac{1}{6.25}(6.25)x = \frac{1}{6.25}(27)$$
$$x = 4.32$$

   It will take Parker 4.32 hours to run 27 miles.

3. Jared walks from baseball practice to his aunt's house, a distance of 6 miles, in 90 minutes. Assuming he walks at a constant speed, C, how far does he walk in 20 minutes? Round your answer to the hundredths place.

   Let y represent the number of miles walked.

   Since $\frac{6}{90} = C$ and $\frac{y}{20} = C$, then

$$\frac{6}{90} = \frac{y}{20}$$
$$6(20) = 90y$$
$$120 = 90y$$
$$\frac{1}{90}(120) = \frac{1}{90}(90)y$$
$$\frac{120}{90} = y$$
$$1.33 \approx y$$

   Jared walks approximately 1.33 miles in 20 minutes.

4. Sammy bikes 3 miles every night for exercise. It takes him exactly 1.75 hours to finish his ride.

a. Assuming he rides at a constant rate, write an equation that represents how many miles, y, Sammy can ride in x hours.

$$\frac{3}{1.75} = \frac{y}{x}$$
$$3x = 1.75y$$
$$\frac{1}{1.75}(3)x = \frac{1}{1.75}(1.75)y$$
$$\frac{3}{1.75}x = y$$

I don't need to define my variables for this problem because they have already done it in the problem.

b. Use your equation from part (a) to complete the table below. Use a calculator, and round all values to the hundredths place.

x (hours)	Linear Equation in y: $\frac{3}{1.75}x = y$	y (miles)
0.25	$\frac{3}{1.75}(0.25) = y$	**0.43**
0.5	$\frac{3}{1.75}(0.5) = y$	**0.86**
0.75	$\frac{3}{1.75}(0.75) = y$	**1.29**
1	$\frac{3}{1.75}(1) = y$	**1.71**
3	$\frac{3}{1.75}(3) = y$	**5.14**

1. Eman walks from the store to her friend's house, 2 miles away. It takes her 35 minutes.
   a. What fraction represents her constant speed, C?
   b. Write the fraction that represents her constant speed, C, if she walks y miles in 10 minutes.
   c. Write and solve a proportion using the fractions from parts (a) and (b) to determine how many miles she walks after 10 minutes. Round your answer to the hundredths place.
   d. Write a two-variable equation to represent how many miles Eman can walk over any time interval.

2. Erika drives from school to soccer practice 1.3 miles away. It takes her 7 minutes.
   a. What fraction represents her constant speed, C?
   b. What fraction represents her constant speed, C, if it takes her x minutes to drive exactly 1 mile?
   c. Write and solve a proportion using the fractions from parts (a) and (b) to determine how much time it takes her to drive exactly 1 mile. Round your answer to the tenths place.
   d. Write a two-variable equation to represent how many miles Erika can drive over any time interval.

3. Darla drives at a constant speed of 45 miles per hour.
   a. If she drives for y miles and it takes her x hours, write the two-variable equation to represent the number of miles Darla can drive in x hours.
   b. Darla plans to drive to the market 14 miles from her house, then to the post office 3 miles from the market, and then return home, which is 15 miles from the post office. Assuming she drives at a constant speed the entire time, how much time will she spend driving as she runs her errands? Round your answer to the hundredths place.

4. Aaron walks from his sister's house to his cousin's house, a distance of 4 miles, in 80 minutes. How far does he walk in 30 minutes?

5. Carlos walks 4 miles every night for exercise. It takes him exactly 63 minutes to finish his walk.
   a. Assuming he walks at a constant rate, write an equation that represents how many miles, y, Carlos can walk in x minutes.
   b. Use your equation from part (a) to complete the table below. Use a calculator, and round all values to the hundredths place.

x (minutes)	Linear Equation:	y (miles)
15		
30		
40		
60		
75		

Example 1

Pauline mows a lawn at a constant rate. Suppose she mows a 35-square-foot lawn in 2.5 minutes. What area, in square feet, can she mow in 10 minutes? t minutes?

t (time in minutes)	Linear Equation:	y(area in square feet)

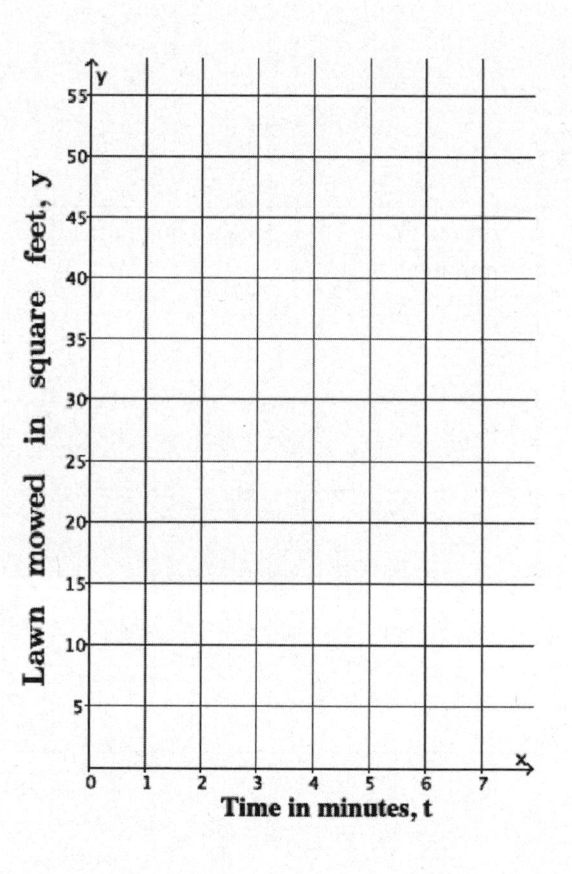

Example 2

Water flows at a constant rate out of a faucet. Suppose the volume of water that comes out in three minutes is 10.5 gallons. How many gallons of water come out of the faucet in t minutes?

t (time in minutes)	Linear Equation:	V (in gallons)
0		
1		
2		
3		
4		

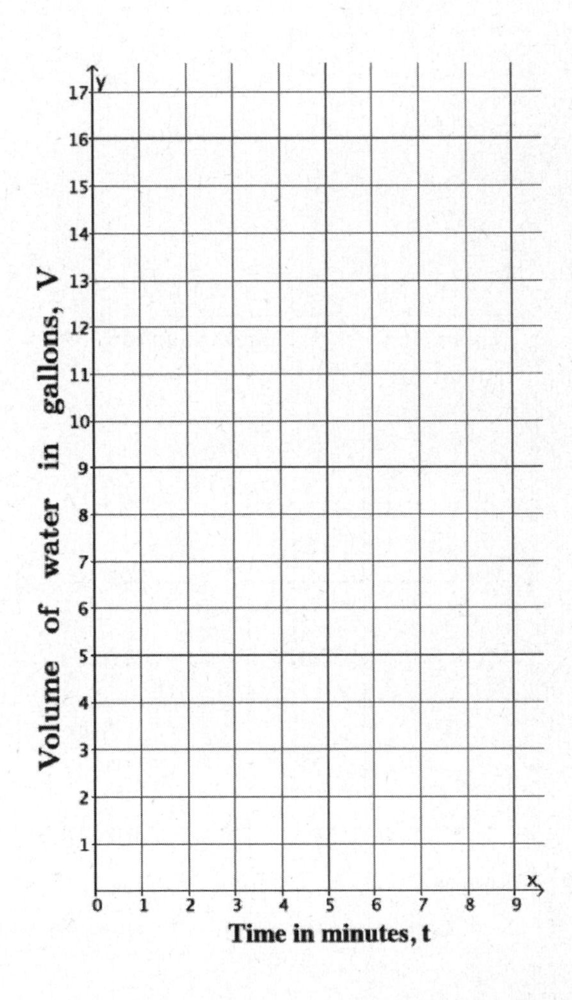

Exercises

1. Juan types at a constant rate. He can type a full page of text in $3\frac{1}{2}$ minutes. We want to know how many pages, p, Juan can type after t minutes.

   a. Write the linear equation in two variables that represents the number of pages Juan types in any given time interval.

   b. Complete the table below. Use a calculator, and round your answers to the tenths place.

t (time in minutes)	Linear Equation:	p (pages typed)
0		
5		
10		
15		
20		

   c. Graph the data on a coordinate plane.

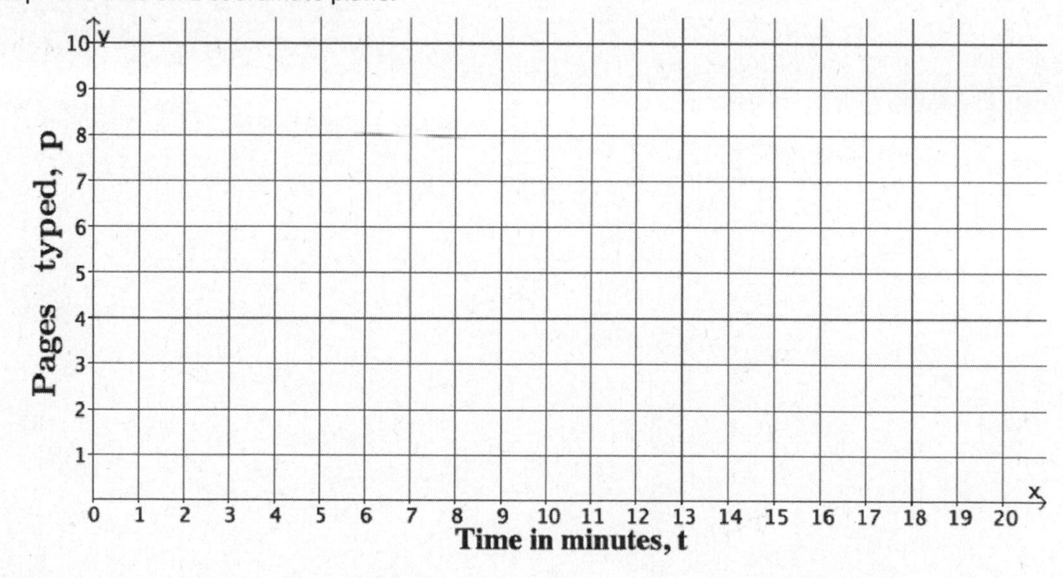

d. About how long would it take Juan to type a 5-page paper? Explain.

2. Emily paints at a constant rate. She can paint 32 square feet in 5 minutes. What area, A, in square feet, can she paint in t minutes?

a. Write the linear equation in two variables that represents the number of square feet Emily can paint in any given time interval.

b. Complete the table below. Use a calculator, and round answers to the tenths place.

t (time in minutes)	Linear Equation:	A (area painted in square feet)
0		
1		
2		
3		
4		

c. Graph the data on a coordinate plane.

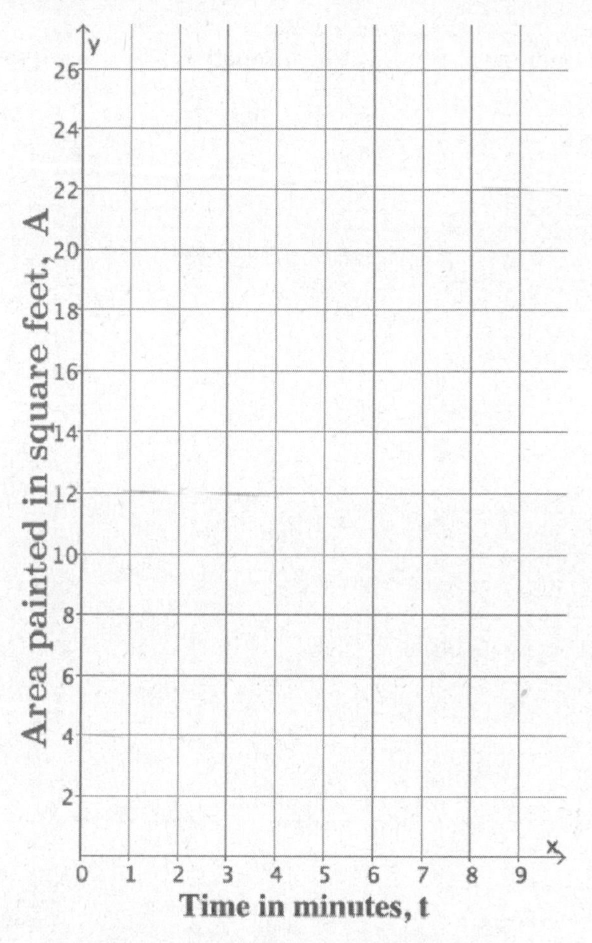

d. About how many square feet can Emily paint in $2\frac{1}{2}$ minutes? Explain.

3. Joseph walks at a constant speed. He walked to a store that is one-half mile away in 6 minutes. How many miles, m, can he walk in t minutes?

a. Write the linear equation in two variables that represents the number of miles Joseph can walk in any given time interval, t.

b. Complete the table below. Use a calculator, and round answers to the tenths place.

t (time in minutes)	Linear Equation:	m (distance in miles)
0		
30		
60		
90		
120		

c. Graph the data on a coordinate plane.

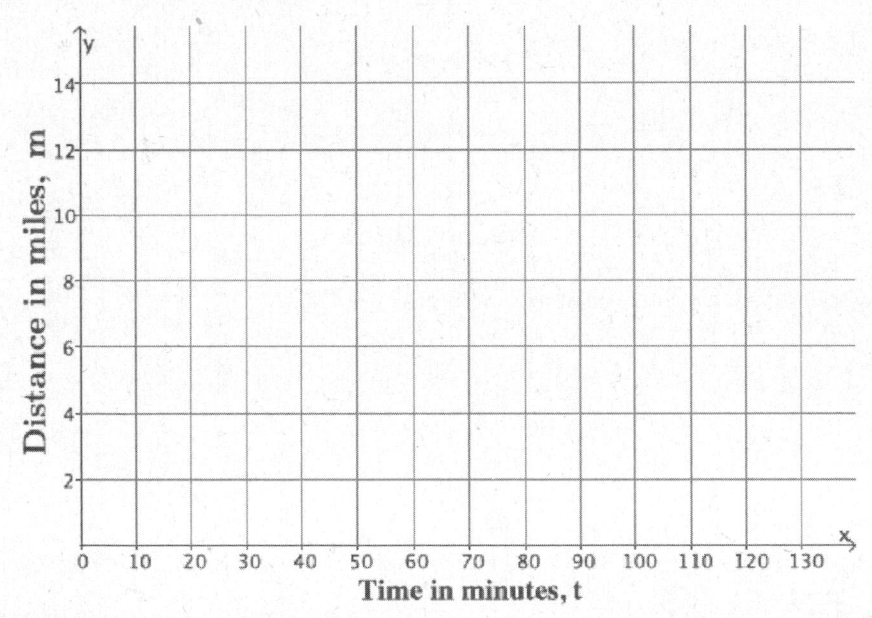

d. Joseph's friend lives 4 miles away from him. About how long would it take Joseph to walk to his friend's house? Explain.

Lesson Summary

When constant rate is stated for a given problem, then you can express the situation as a two-variable equation. The equation can be used to complete a table of values that can then be graphed on a coordinate plane.

Name ____________________ Date ____________

Vicky reads at a constant rate. She can read 5 pages in 9 minutes. We want to know how many pages, p, Vicky can read after t minutes.

a. Write a linear equation in two variables that represents the number of pages Vicky reads in any given time interval.

b. Complete the table below. Use a calculator, and round answers to the tenths place.

t (time in minutes)	Linear Equation:	p (pages read)
0		
20		
40		
60		

c. About how long would it take Vicky to read 25 pages? Explain.

1. A bus travels at a constant rate of 40 miles per hour.

   What is the distance, d, in miles, that the bus travels in t hours?

   Let C be the constant rate the bus travels. Then, $\frac{40}{1} = C$, and $\frac{d}{t} = C$; therefore, $\frac{40}{1} = \frac{d}{t}$

   $$\frac{40}{1} = \frac{d}{t}$$
   $$d = 40t$$

   If I can write two fractions, each equal to the constant rate, C, then I can use the proportional relationship to solve for d.

2. A teenage boy named Harry can consume 8 hot dogs in 1.25 hours. Assume that the young man eats at a constant rate.

   a. How many hot dogs, y, can be consumed by Harry in t hours?

   Let C be the constant rate Harry eats hot dogs. Then, $\frac{8}{1.25} = C$, and $\frac{y}{t} = C$; therefore, $\frac{8}{1.25} = \frac{y}{t}$.

   $$\frac{8}{1.25} = \frac{y}{t}$$
   $$1.25y = 8t$$
   $$\frac{1.25}{1.25}y = \frac{8}{1.25}t$$
   $$y = 6.4t$$

   b. Pretend that he can eat every hour of every day for a week. How many hot dogs would Harry consume?

   24 hours a day for 7 days is a total of 168 hours.

   $$y = 6.4t$$
   $$y = 6.4(168)$$
   $$y = 1{,}075.2$$

   Once I figure out how many hours are in a week, I can use my equation from part (a) to determine the answer.

   Harry would consume about 1,075 hot dogs in one week.

3. Your cell phone company charges at a constant rate. The company charges $1.00 for 4 minutes of use.

   a. Write an equation to represent the number of dollars, d, that will be charged over any time interval, t.

   Let C be the constant rate charged per minute. Then, $\frac{1.00}{4} = C$, and $\frac{d}{t} = C$; therefore, $\frac{1.00}{4} = \frac{d}{t}$.

$$\frac{1}{4} = \frac{d}{t}$$
$$4d = 1t$$
$$\frac{1}{4}(4)d = \frac{1}{4}(1)t$$
$$d = 0.25t$$

   b. Complete the table below.

t (time in minutes)	Linear Equation: $d = 0.25t$	d (cost in dollars)
0	$d = 0.25(0)$	**0**
5	$d = 0.25(5)$	**1.25**
10	$d = 0.25(10)$	**2.50**
15	$d = 0.25(15)$	**3.75**
20	$d = 0.25(20)$	**5.00**

   c. Graph the data as points on a coordinate plane.

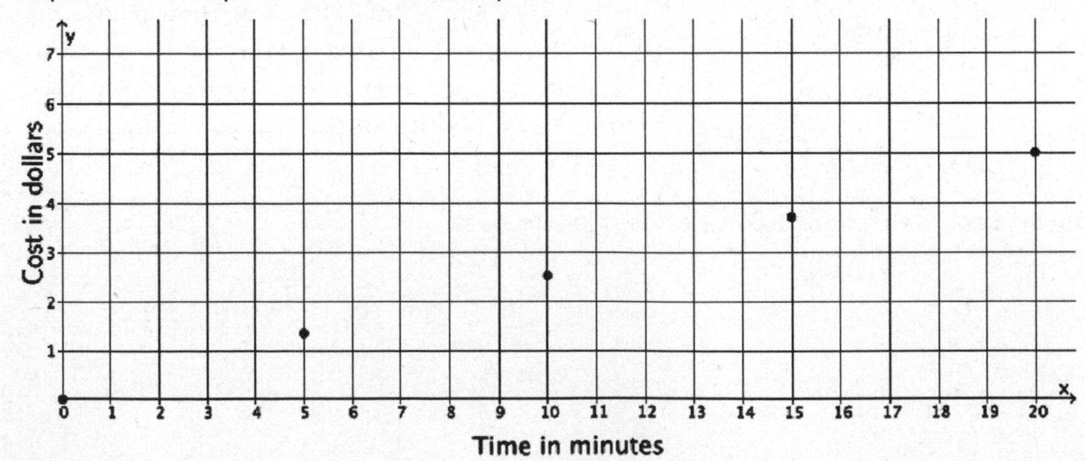

d. You used your phone for 18 minutes. About how much will your bill be? Explain.

It will cost between $\$3.75$ and $\$5$. I located 18 on the x-axis because that is the number of minutes I used. That x-value is between the known costs for 15 minutes and 20 minutes. So my bill will probably be closer to $\$5$ because 18 is closer to 20 than to 15.

1. A train travels at a constant rate of 45 miles per hour.

   a. What is the distance, d, in miles, that the train travels in t hours?

   b. How many miles will it travel in 2.5 hours?

2. Water is leaking from a faucet at a constant rate of $\frac{1}{3}$ gallon per minute.

   a. What is the amount of water, w, in gallons per minute, that is leaked from the faucet after t minutes?

   b. How much water is leaked after an hour?

3. A car can be assembled on an assembly line in 6 hours. Assume that the cars are assembled at a constant rate.

   a. How many cars, y, can be assembled in t hours?

   b. How many cars can be assembled in a week?

4. A copy machine makes copies at a constant rate. The machine can make 80 copies in $2\frac{1}{2}$ minutes.

   a. Write an equation to represent the number of copies, n, that can be made over any time interval in minutes, t.

   b. Complete the table below.

***t* (time in minutes)**	**Linear Equation:**	***n* (number of copies)**
0		
0.25		
0.5		
0.75		
1		

c. Graph the data on a coordinate plane.

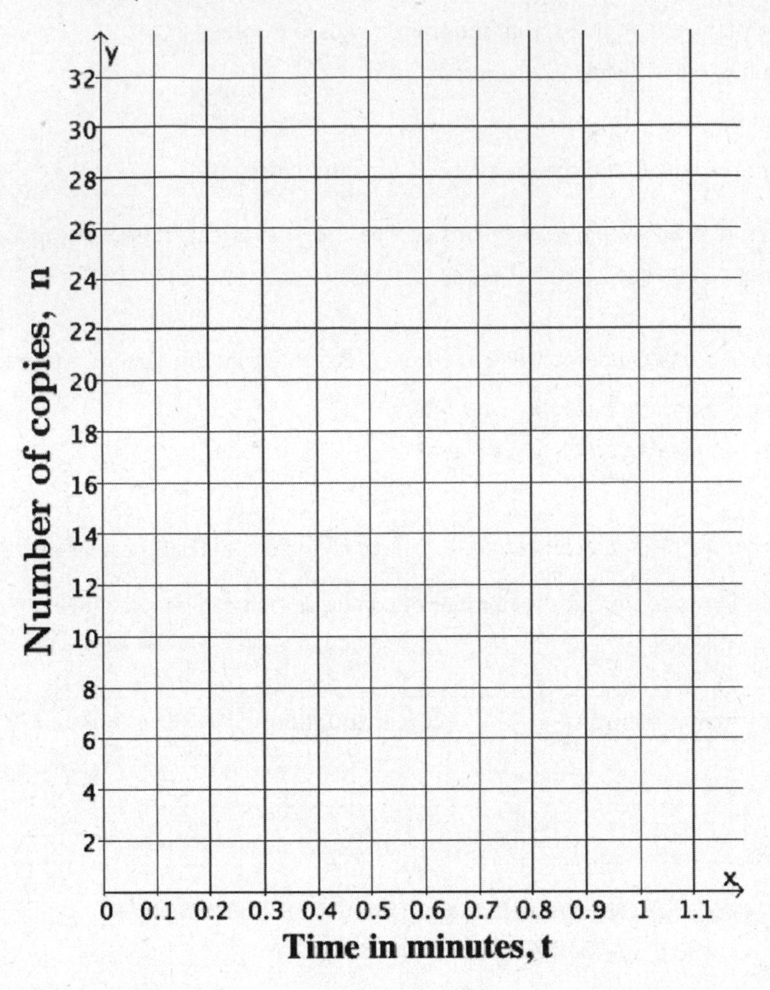

d. The copy machine runs for 20 seconds and then jams. About how many copies were made before the jam occurred? Explain.

5. Connor runs at a constant rate. It takes him 34 minutes to run 4 miles.

   a. Write the linear equation in two variables that represents the number of miles Connor can run in any given time interval in minutes, t.

   b. Complete the table below. Use a calculator, and round answers to the tenths place.

t (time in minutes)	Linear Equation:	m (distance in miles)
0		
15		
30		
45		
60		

   c. Graph the data on a coordinate plane.

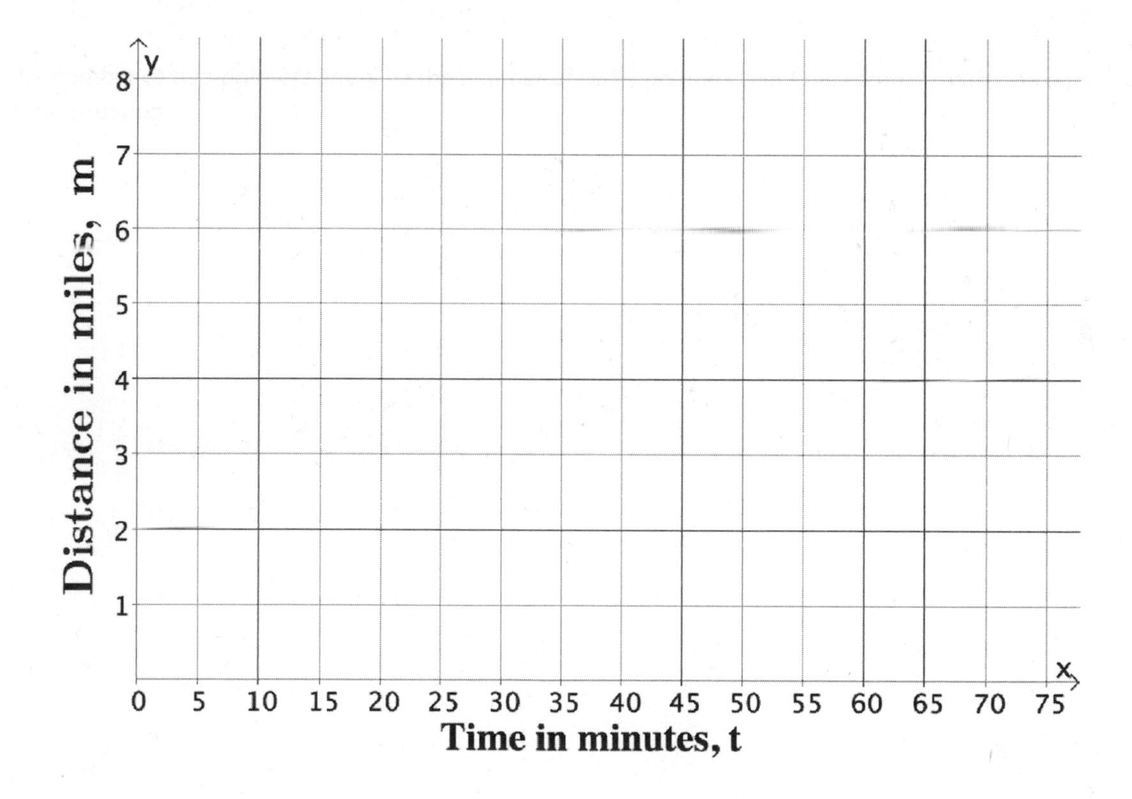

   d. Connor ran for 40 minutes before tripping and spraining his ankle. About how many miles did he run before he had to stop? Explain.

Exploratory Challenge/Exercises

1. Find five solutions for the linear equation $x + y = 3$, and plot the solutions as points on a coordinate plane.

x	Linear Equation: $x + y = 3$	y

2. Find five solutions for the linear equation $2x - y = 10$, and plot the solutions as points on a coordinate plane.

x	Linear Equation: $2x - y = 10$	y

3. Find five solutions for the linear equation $x + 5y = 21$, and plot the solutions as points on a coordinate plane.

x	Linear Equation: $x + 5y = 21$	y

4. Consider the linear equation $\frac{2}{5}x + y = 11$.

   a. Will you choose to fix values for x or y? Explain.

   b. Are there specific numbers that would make your computational work easier? Explain.

c. Find five solutions to the linear equation $\frac{2}{5}x + y = 11$, and plot the solutions as points on a coordinate plane.

x	**Linear Equation:** $\frac{2}{5}x + y = 11$	y

5. At the store, you see that you can buy a bag of candy for $2 and a drink for $1. Assume you have a total of $35 to spend. You are feeling generous and want to buy some snacks for you and your friends.

   a. Write an equation in standard form to represent the number of bags of candy, x, and the number of drinks, y, that you can buy with $35.

b. Find five solutions to the linear equation from part (a), and plot the solutions as points on a coordinate plane.

x	**Linear Equation:**	y

Lesson Summary

A linear equation in two-variables x and y is in standard form if it is of the form $ax + by = c$ for numbers a, b, and c, where a and b are both not zero. The numbers a, b, and c are called constants.

A solution to a linear equation in two variables is the ordered pair (x,y) that makes the given equation true. Solutions can be found by fixing a number for x and solving for y or fixing a number for y and solving for x.

Name ______________________________ Date ______________

1. Is the point $(1,3)$ a solution to the linear equation $5x - 9y = 32$? Explain.

2. Find three solutions for the linear equation $4x - 3y = 1$, and plot the solutions as points on a coordinate plane.

x	**Linear Equation:** $4x - 3y = 1$	y

1. Consider the linear equation $x - \frac{2}{5}y = 4$.

   a. Will you choose to fix values for x or y? Explain.

   If I fix values for y, it will make the computations easier. Solving for x can be done in one step.

   b. Are there specific numbers that would make your computational work easier? Explain.

   Values for y that are multiples of 5 will make the computations easier. When I multiply $\frac{2}{5}$ by a multiple of 5, I will get a whole number.

   c. Find three solutions to the linear equation $x - \frac{2}{5}y = 4$, and plot the solutions as points on a coordinate plane.

I'll use the numbers 5, 10, and 15 for y in my table. Once substituted into the equation, I'll get the values for x. Then each pair of x and y will be a point on my graph.

x	Linear Equation: $x - \frac{2}{5}y = 4$	y
6	$x - \frac{2}{5}(5) = 4$ $x - 2 = 4$ $x = 6$	5
8	$x - \frac{2}{5}(10) = 4$ $x - 4 = 4$ $x = 8$	10
10	$x - \frac{2}{5}(15) = 4$ $x - 6 = 4$ $x = 10$	15

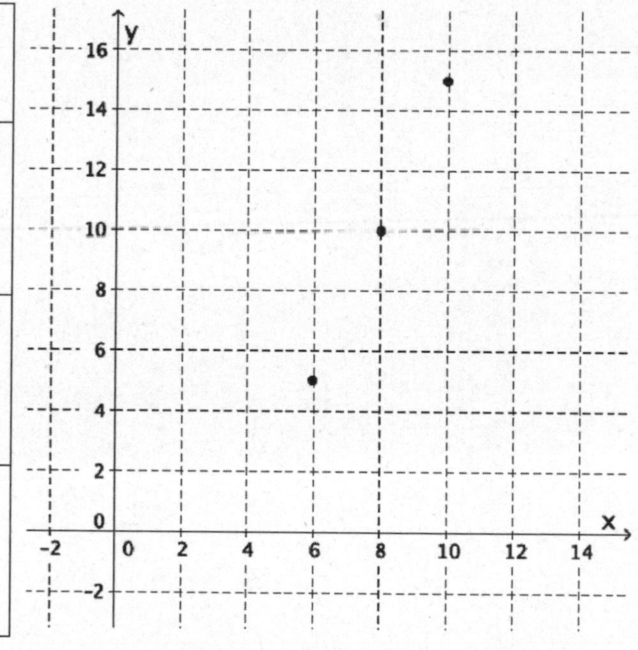

1. Consider the linear equation $x - \frac{3}{2}y = -2$.
   a. Will you choose to fix values for x or y? Explain.
   b. Are there specific numbers that would make your computational work easier? Explain.
   c. Find five solutions to the linear equation $x - \frac{3}{2}y = -2$, and plot the solutions as points on a coordinate plane.

x	Linear Equation: $x - \frac{3}{2}y = -2$	y

2. Find five solutions for the linear equation $\frac{1}{3}x + y = 12$, and plot the solutions as points on a coordinate plane.

3. Find five solutions for the linear equation $-x + \frac{3}{4}y = -6$, and plot the solutions as points on a coordinate plane.

4. Find five solutions for the linear equation $2x + y = 5$, and plot the solutions as points on a coordinate plane.

5. Find five solutions for the linear equation $3x - 5y = 15$, and plot the solutions as points on a coordinate plane.

Exercises

1. Find at least ten solutions to the linear equation $3x + y = -8$, and plot the points on a coordinate plane.

x	**Linear Equation:** $3x + y = -8$	y

What shape is the graph of the linear equation taking?

2. Find at least ten solutions to the linear equation $x - 5y = 11$, and plot the points on a coordinate plane.

x	**Linear Equation:** $x - 5y = 11$	y

What shape is the graph of the linear equation taking?

3. Compare the solutions you found in Exercise 1 with a partner. Add the partner's solutions to your graph.

   Is the prediction you made about the shape of the graph still true? Explain.

4. Compare the solutions you found in Exercise 2 with a partner. Add the partner's solutions to your graph.

   Is the prediction you made about the shape of the graph still true? Explain.

5. Joey predicts that the graph of $-x + 2y = 3$ will look like the graph shown below. Do you agree? Explain why or why not.

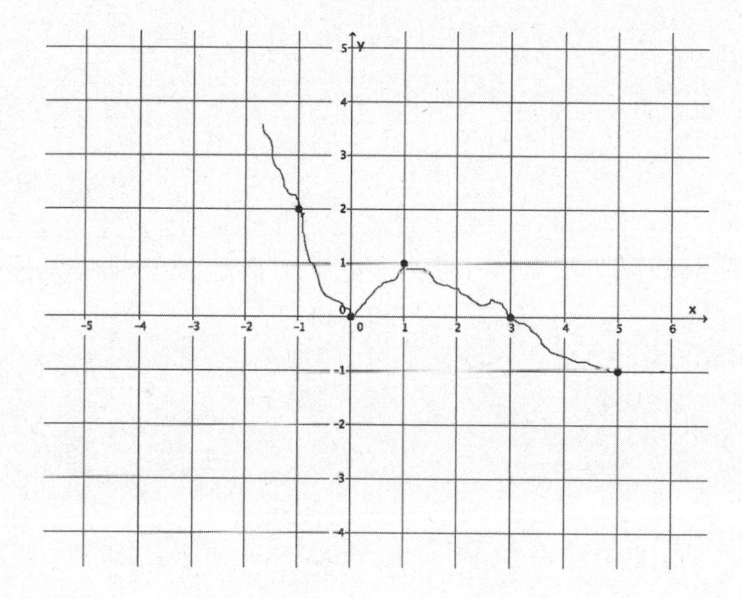

6. We have looked at some equations that appear to be lines. Can you write an equation that has solutions that do not form a line? Try to come up with one, and prove your assertion on the coordinate plane.

Lesson Summary

One way to determine if a given point is on the graph of a linear equation is by checking to see if it is a solution to the equation. Note that all graphs of linear equations appear to be lines.

Name ______________________________ Date ________________

1. Ethan found solutions to the linear equation $3x - y = 8$ and graphed them. What shape is the graph of the linear equation taking?

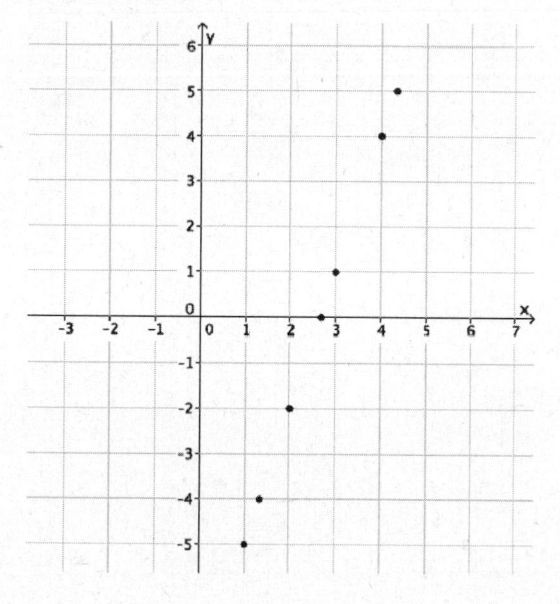

2. Could the following points be on the graph of $-x + 2y = 5$?

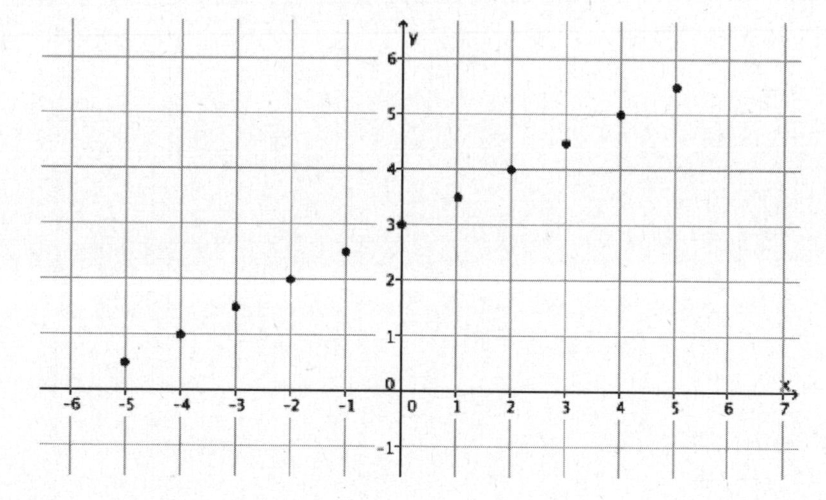

1. Find at least five solutions to the linear equation $\frac{1}{4}x + y = 7$, and plot the points on a coordinate plane. What shape is the graph of the linear equation taking?

> I should choose values for x that are multiples of 4. I should also be sure to select some positive values for x as well as negative.

x	$\frac{1}{4}x + y = 7$	y
-8	$\frac{1}{4}(-8) + y = 7$ $-2 + y = 7$ $-2 + 2 + y = 7 + 2$ $y = 9$	9
-4	$\frac{1}{4}(-4) + y = 7$ $-1 + y = 7$ $-1 + 1 + y = 7 + 1$ $y = 8$	8
0	$\frac{1}{4}(0) + y = 7$ $0 + y = 7$ $y = 7$	7
4	$\frac{1}{4}(4) + y = 7$ $1 + y = 7$ $1 - 1 + y = 7 - 1$ $y = 6$	6
8	$\frac{1}{4}(8) + y = 7$ $2 + y = 7$ $2 - 2 + y = 7 - 2$ $y = 5$	5

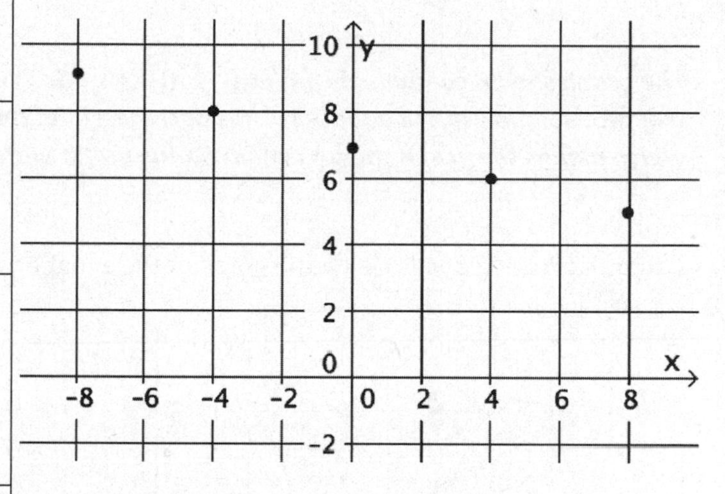

2. Can the following points be on the graph of the equation $x - 3y = 3$? Explain

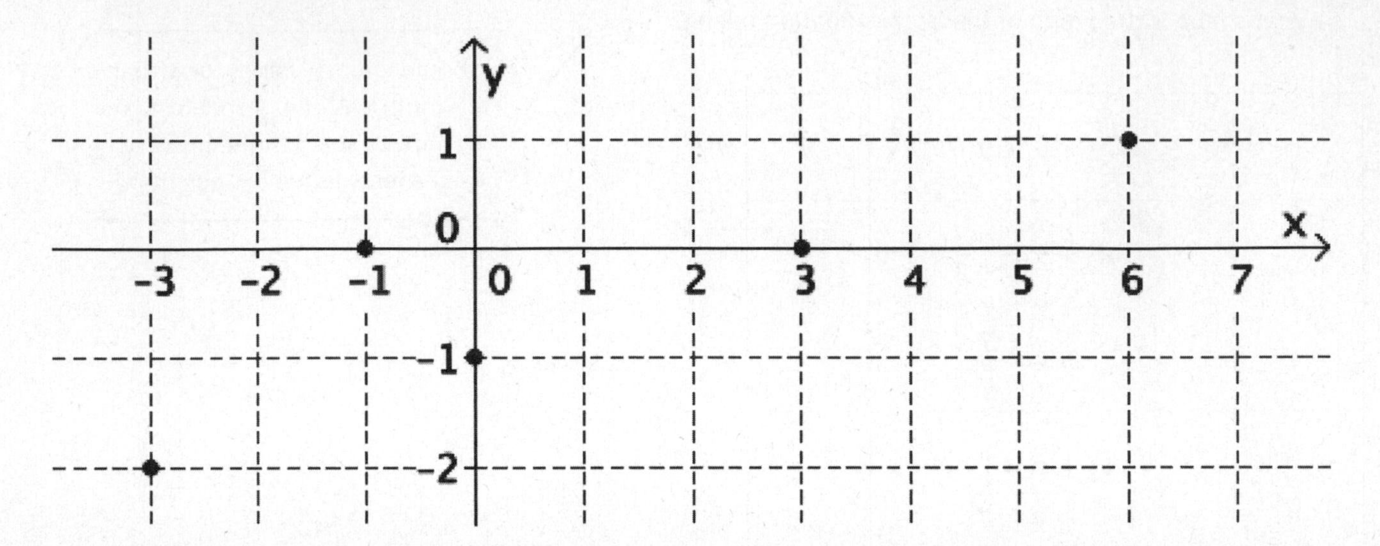

The graph shown contains the point $(-1, 0)$. If $(-1, 0)$ is on the graph of the linear equation, then it will be a solution to the equation. It is not; therefore, the point cannot be on the graph of the equation, which means the graph shown cannot be the graph of the equation $x - 3y = 3$.

3. Can the following points be on the graph of the equation $2x + 4y = 6$? Explain

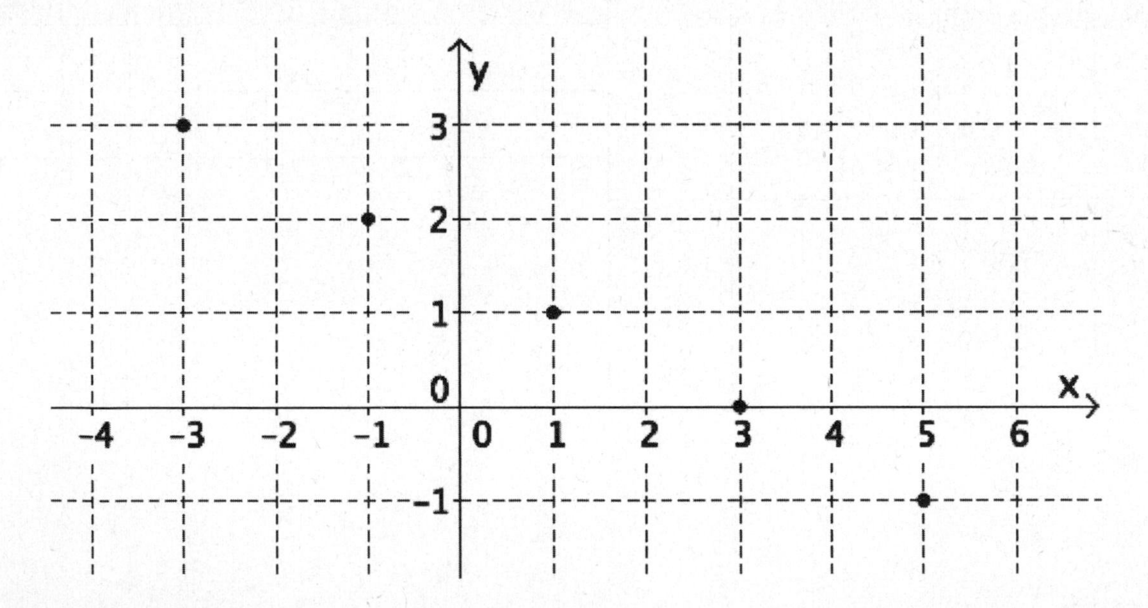

Yes, this graph is of the equation $2x + 4y = 6$ because each point on the graph represents a solution to the linear equation $2x + 4y = 6$.

1. Find at least ten solutions to the linear equation $\frac{1}{2}x + y = 5$, and plot the points on a coordinate plane.

   What shape is the graph of the linear equation taking?

2. Can the following points be on the graph of the equation $x - y = 0$? Explain.

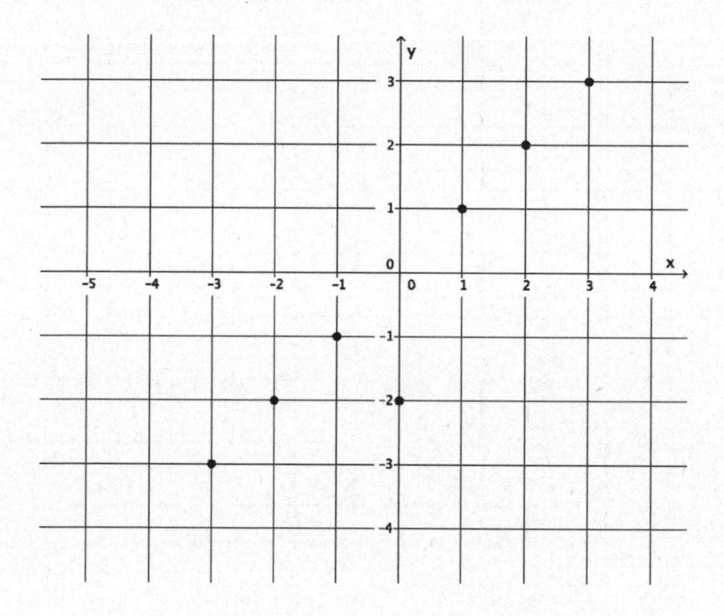

3. Can the following points be on the graph of the equation $x + 2y = 2$? Explain.

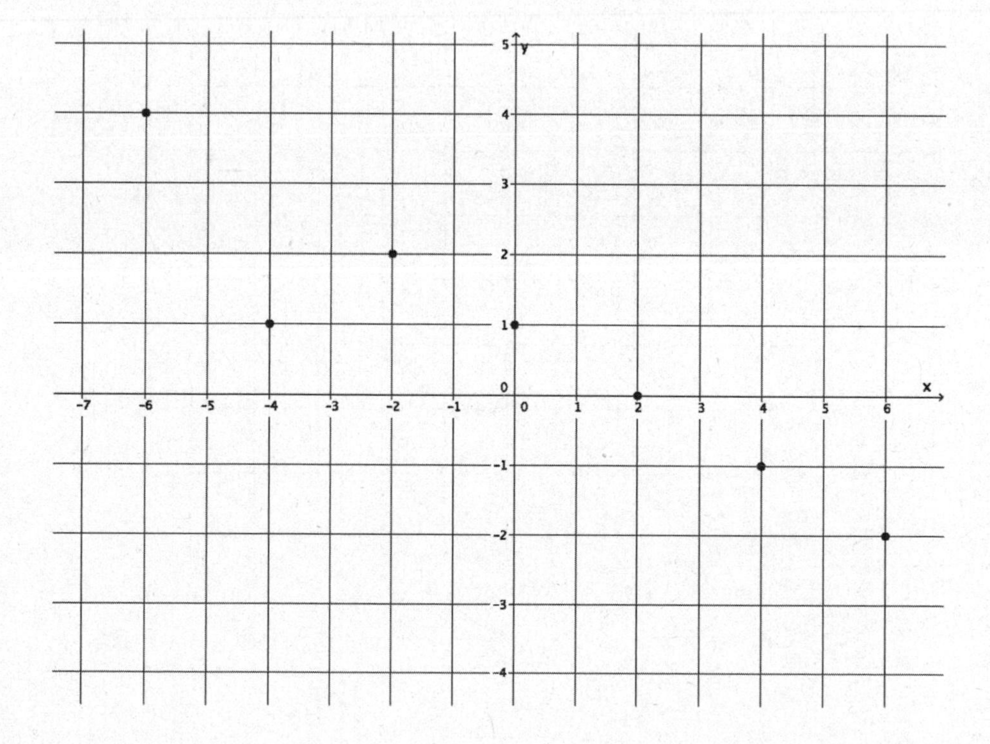

4. Can the following points be on the graph of the equation $x - y = 7$? Explain.

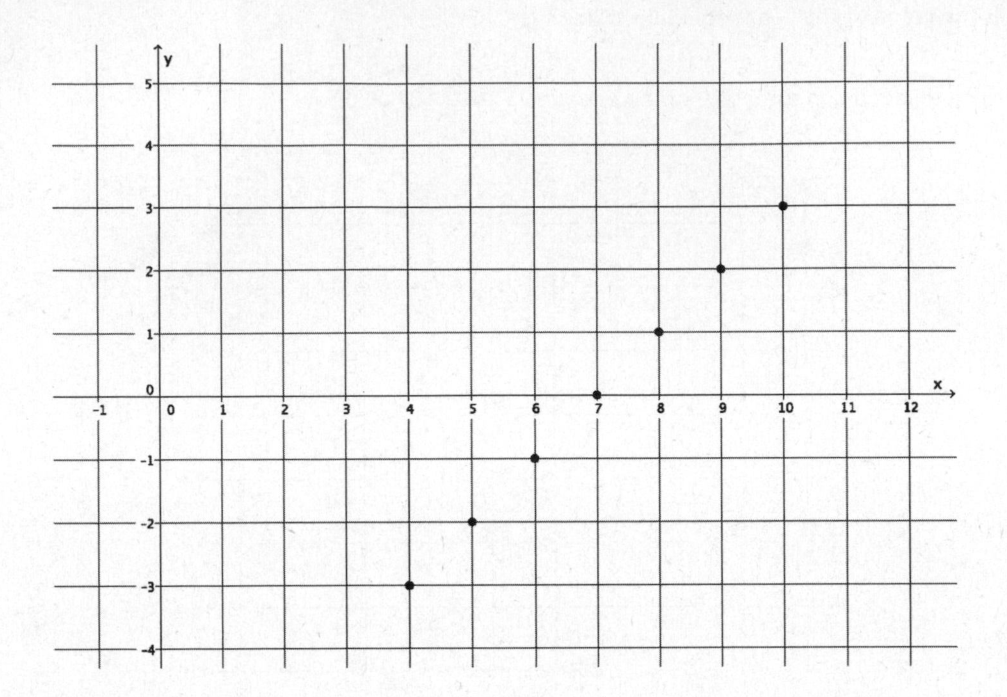

5. Can the following points be on the graph of the equation $x + y = 2$? Explain.

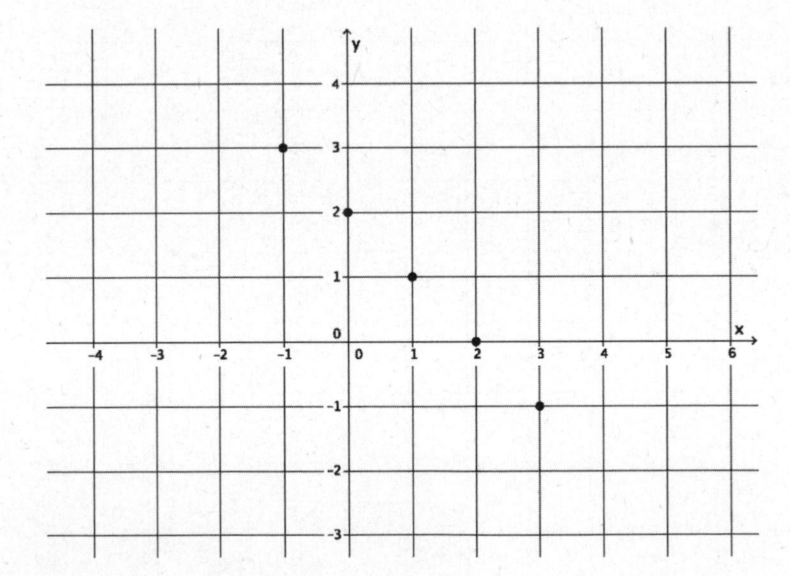

6. Can the following points be on the graph of the equation $2x - y = 9$? Explain.

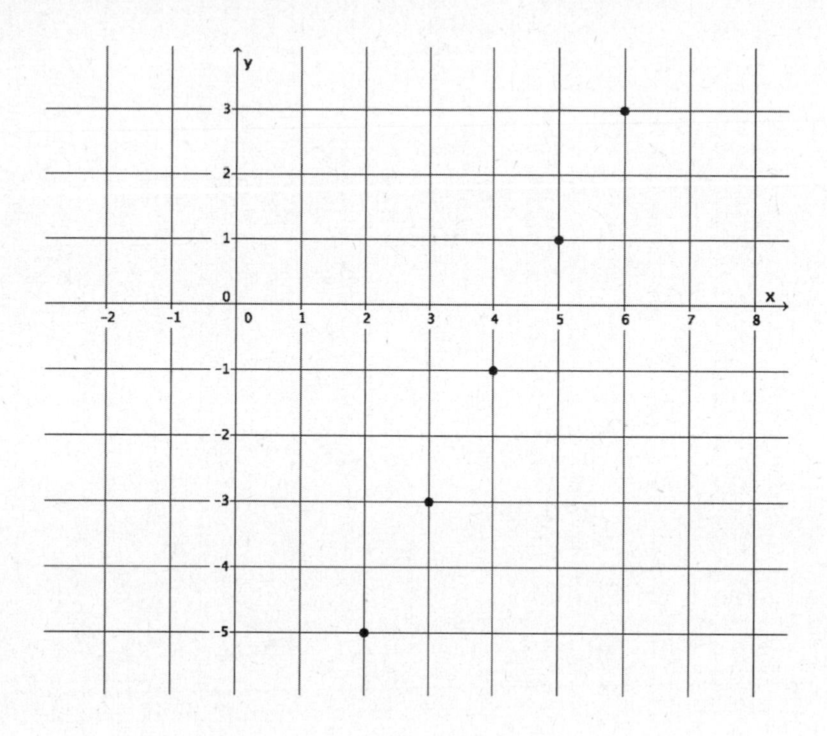

7. Can the following points be on the graph of the equation $x - y = 1$? Explain.

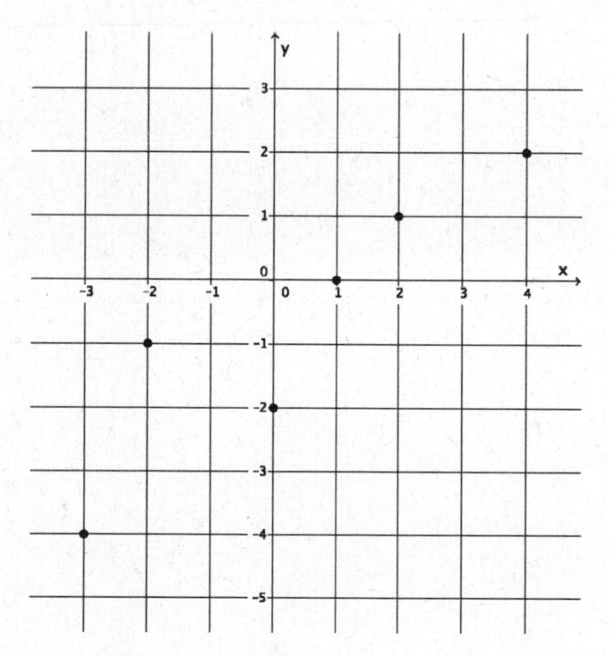

Exercises

1. Find at least four solutions to graph the linear equation $1x + 2y = 5$.

2. Find at least four solutions to graph the linear equation $1x + 0y = 5$.

3. What was different about the equations in Exercises 1 and 2? What effect did this change have on the graph?

4. Graph the linear equation $x = -2$.

5. Graph the linear equation $x = 3$.

6. What will the graph of $x = 0$ look like?

7. Find at least four solutions to graph the linear equation $2x + 1y = 2$.

8. Find at least four solutions to graph the linear equation $0x + 1y = 2$.

9. What was different about the equations in Exercises 7 and 8? What effect did this change have on the graph?

10. Graph the linear equation $y = -2$.

11. Graph the linear equation $y = 3$.

12. What will the graph of $y = 0$ look like?

Lesson Summary

In a coordinate plane with perpendicular x- and y-axes, a *vertical line* is either the y-axis or any other line parallel to the y-axis. The graph of the linear equation in two variables $ax + by = c$, where $a = 1$ and $b = 0$, is the graph of the equation $x = c$. The graph of $x = c$ is the vertical line that passes through the point $(c, 0)$.

In a coordinate plane with perpendicular x- and y-axes, a *horizontal line* is either the x-axis or any other line parallel to the x-axis. The graph of the linear equation in two variables $ax + by = c$, where $a = 0$ and $b = 1$, is the graph of the equation $y = c$. The graph of $y = c$ is the horizontal line that passes through the point $(0, c)$.

Name______________________________ Date ______________

1. Graph the linear equation $ax + by = c$, where $a = 0$, $b = 1$, and $c = 1.5$.

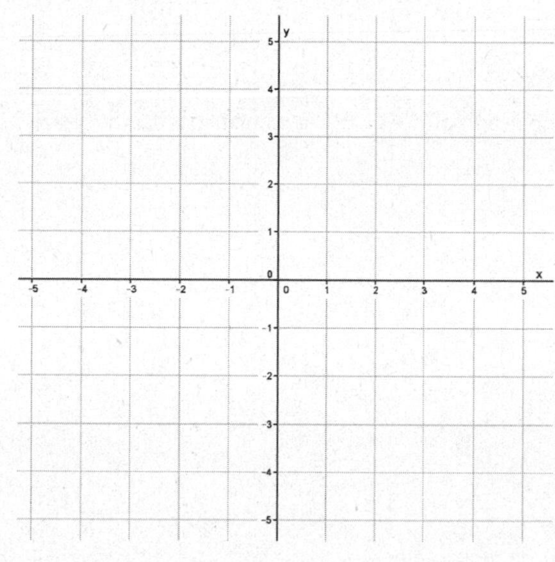

2. Graph the linear equation $ax + by = c$, where $a = 1$, $b = 0$, and $c = -\frac{5}{2}$.

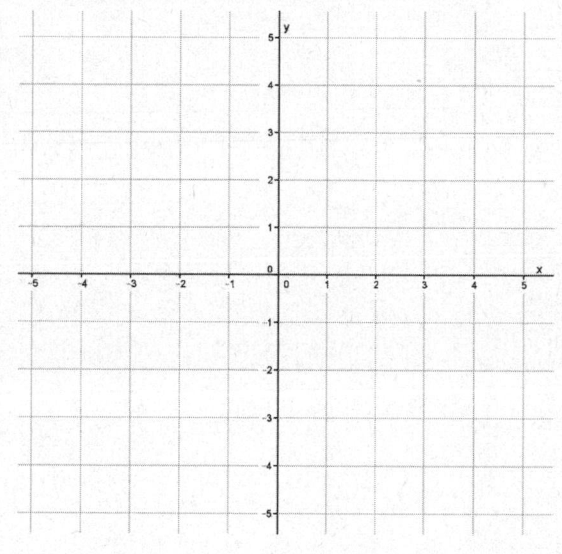

3. What linear equation represents the graph of the line that coincides with the x-axis?

4. What linear equation represents the graph of the line that coincides with the y-axis?

1. Graph the two-variable linear equation $ax + by = c$, where $a = 0$, $b = -2$, and $c = 6$.

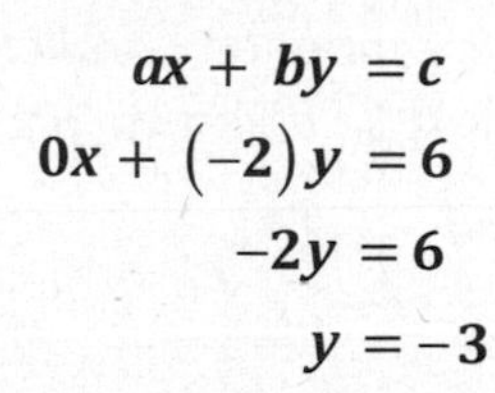

I'm not sure how to graph this, so I'll find some solutions using a table like in the last lesson.

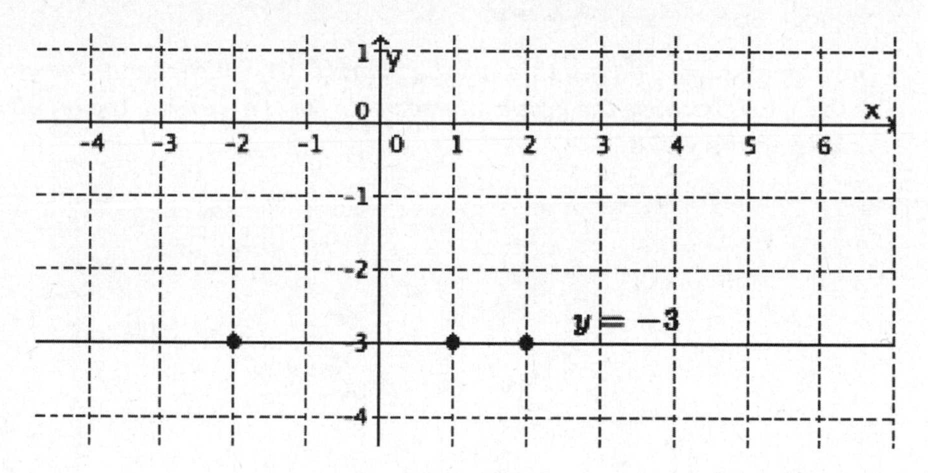

2. Graph the linear equation $x = 1$.

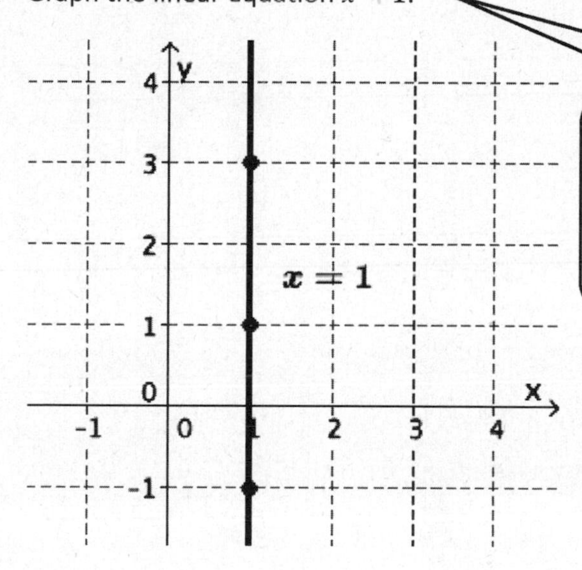

I know that this will either be a horizontal or vertical line. Since the equation is $x = 1$, that means that no matter what value I choose for y, the x-value will always be one.

3. Explain why the graph of a linear equation in the form of $x = c$ is the vertical line, parallel to the y-axis passing through the point $(c, 0)$.

   The graph of $x = c$ passes through the point $(c, 0)$, which means the graph of $x = c$ cannot be parallel to the x-axis because the graph intersects it. For that reason, the graph of $x = c$ must be a vertical line parallel to the y-axis.

1. Graph the two-variable linear equation $ax + by = c$, where $a = 0$, $b = 1$, and $c = -4$.

2. Graph the two-variable linear equation $ax + by = c$, where $a = 1$, $b = 0$, and $c = 9$.

3. Graph the linear equation $y = 7$.

4. Graph the linear equation $x = 1$.

5. Explain why the graph of a linear equation in the form of $y = c$ is the horizontal line, parallel to the x-axis passing through the point $(0, c)$.

6. Explain why there is only one line with the equation $y = c$ that passes through the point $(0, c)$.

Opening Exercise

Graph A

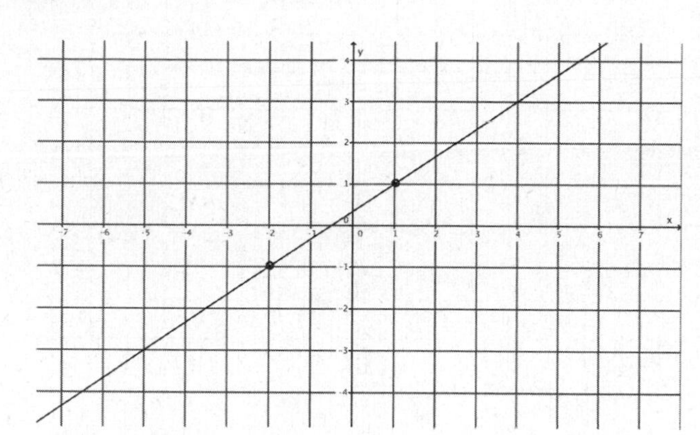

Graph B

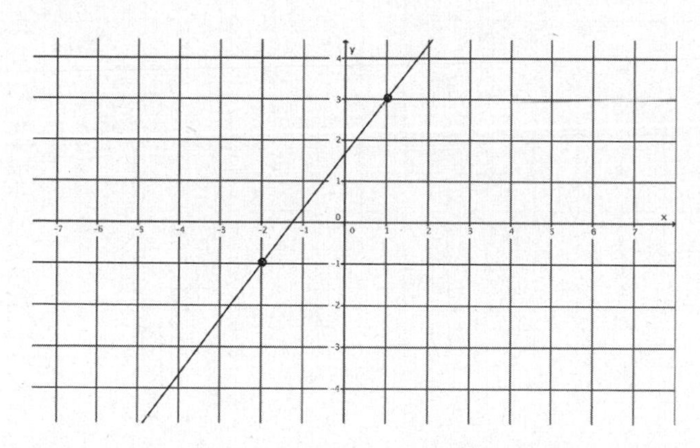

a. Which graph is steeper?

b. Write directions that explain how to move from one point on the graph to the other for both Graph A and Graph B.

c. Write the directions from part (b) as ratios, and then compare the ratios. How does this relate to which graph was steeper in part (a)?

Pair 1:

Graph A

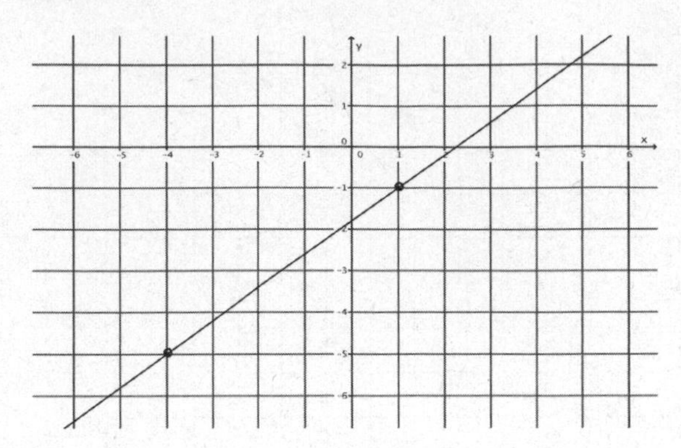

Graph B

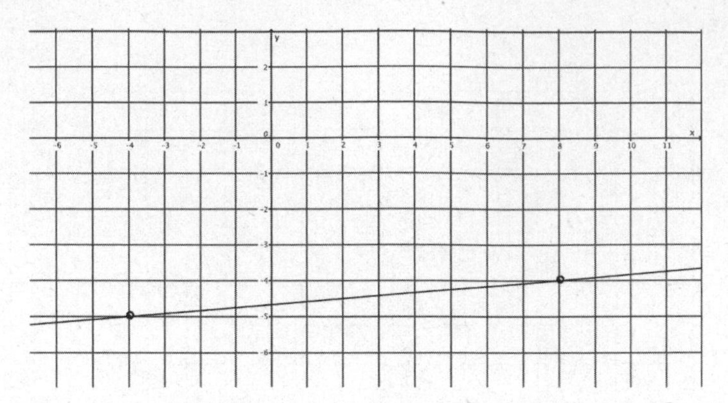

a. Which graph is steeper?

b. Write directions that explain how to move from one point on the graph to the other for both Graph A and Graph B.

c. Write the directions from part (b) as ratios, and then compare the ratios. How does this relate to which graph was steeper in part (a)?

Pair 2:

Graph A

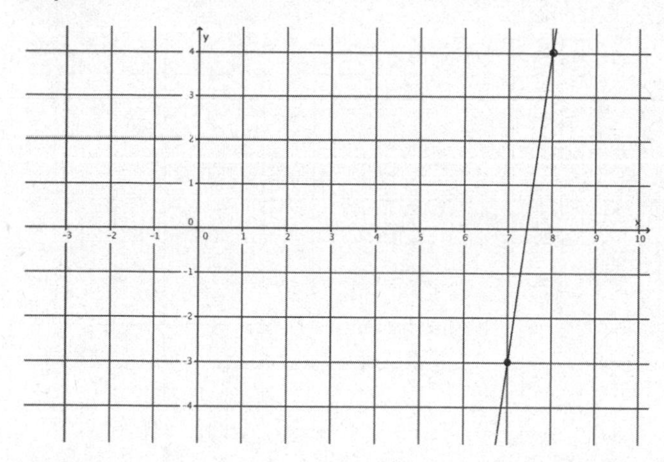

Graph B

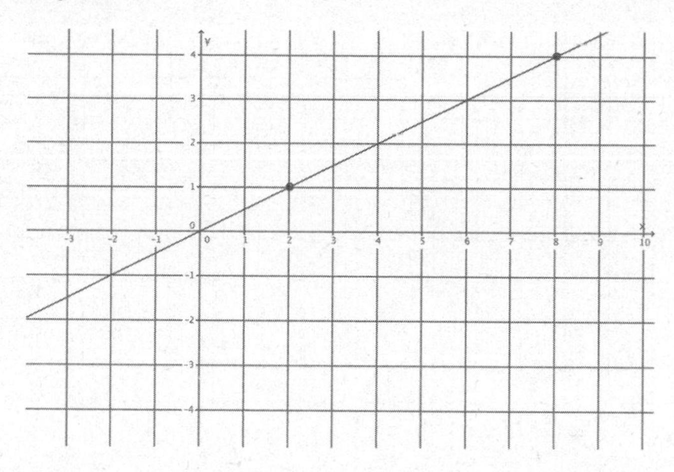

a. Which graph is steeper?

b. Write directions that explain how to move from one point on the graph to the other for both Graph A and Graph B.

c. Write the directions from part (b) as ratios, and then compare the ratios. How does this relate to which graph was steeper in part (a)?

Pair 3:

Graph A

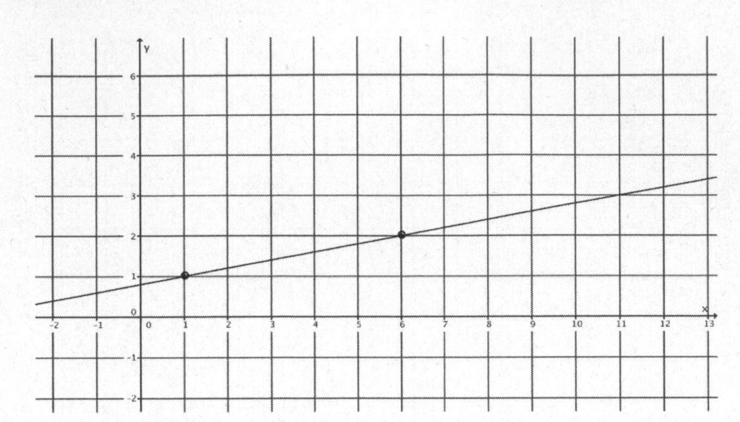

Graph B

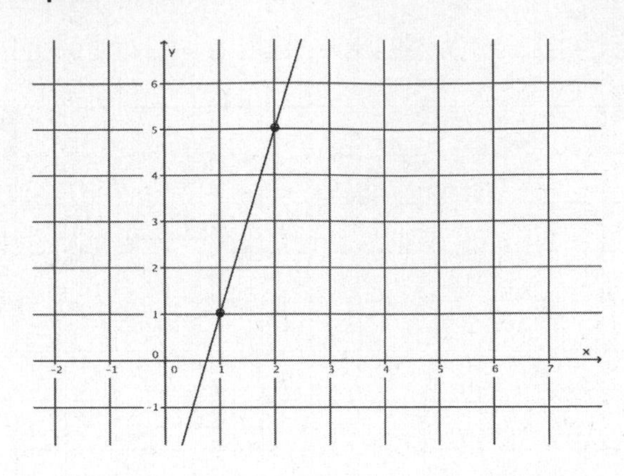

a. Which graph is steeper?

b. Write directions that explain how to move from one point on the graph to the other for both Graph A and Graph B.

c. Write the directions from part (b) as ratios, and then compare the ratios. How does this relate to which graph was steeper in part (a)?

Pair 4:

Graph A

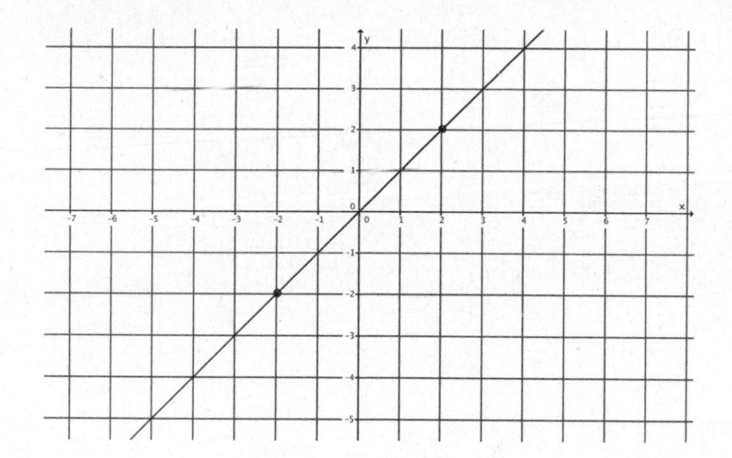

Graph B

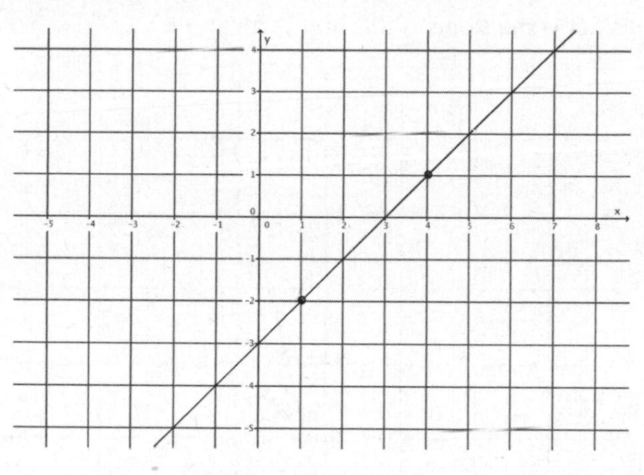

a. Which graph is steeper?

b. Write directions that explain how to move from one point on the graph to the other for both Graph A and Graph B.

c. Write the directions from part (b) as ratios, and then compare the ratios. How does this relate to which graph was steeper in part (a)?

Exercises

Use your transparency to find the slope of each line if needed.

1. What is the slope of this non-vertical line?

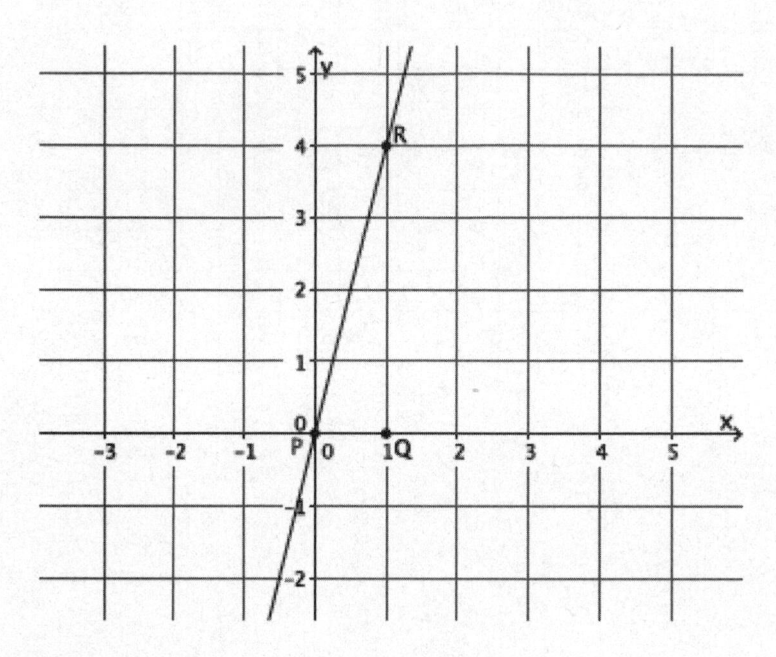

2. What is the slope of this non-vertical line?

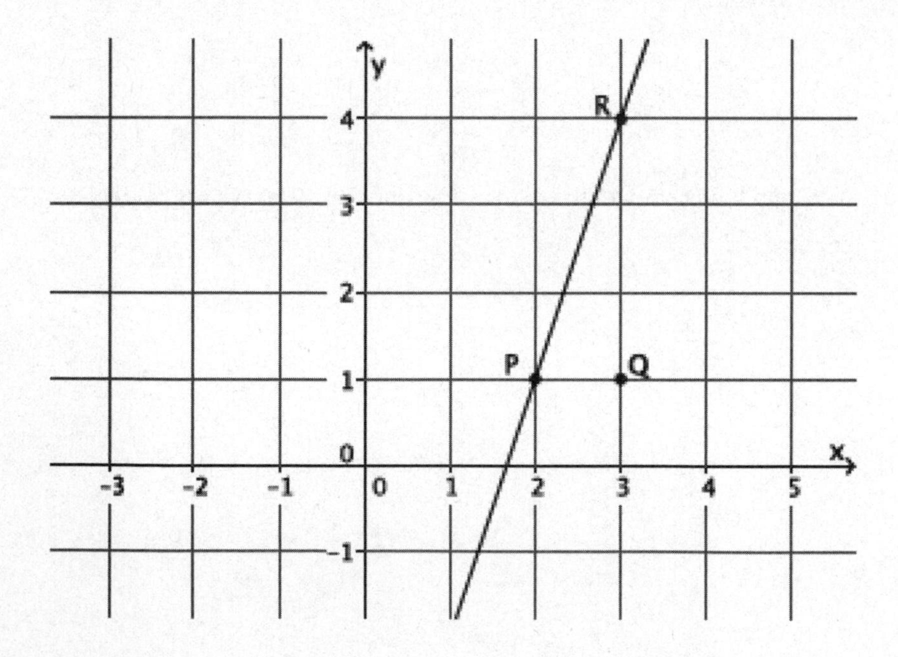

3. Which of the lines in Exercises 1 and 2 is steeper? Compare the slopes of each of the lines. Is there a relationship between steepness and slope?

4. What is the slope of this non-vertical line?

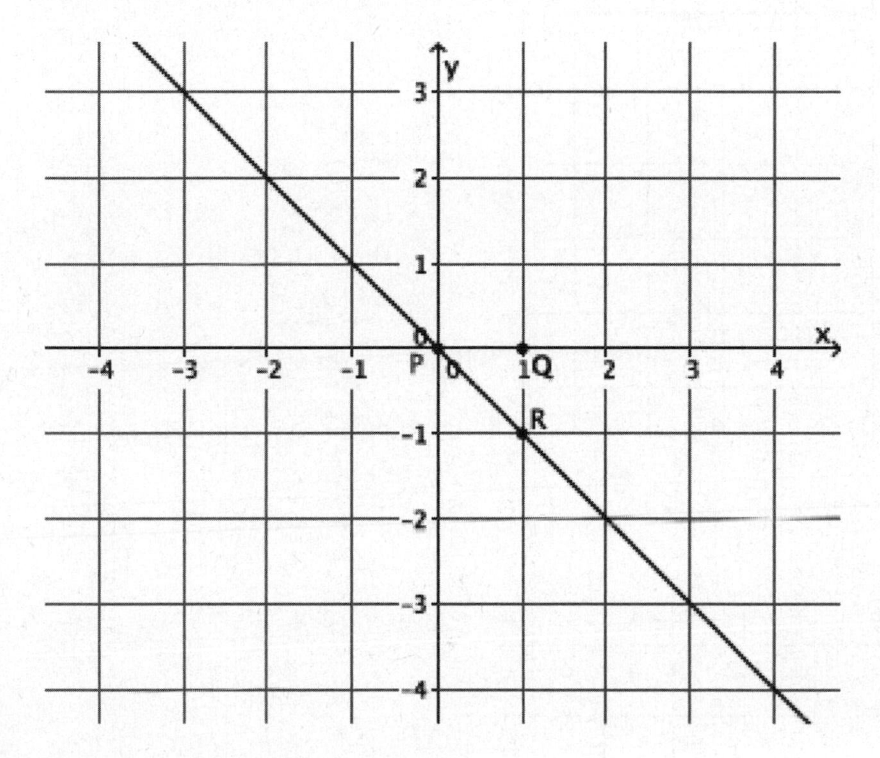

5. What is the slope of this non-vertical line?

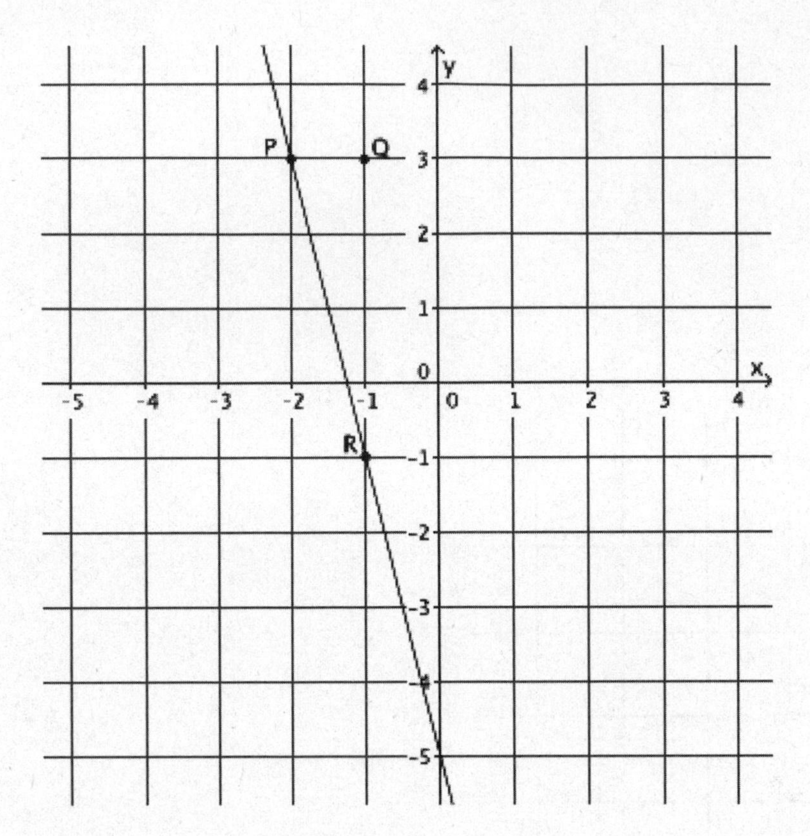

6. What is the slope of this non-vertical line?

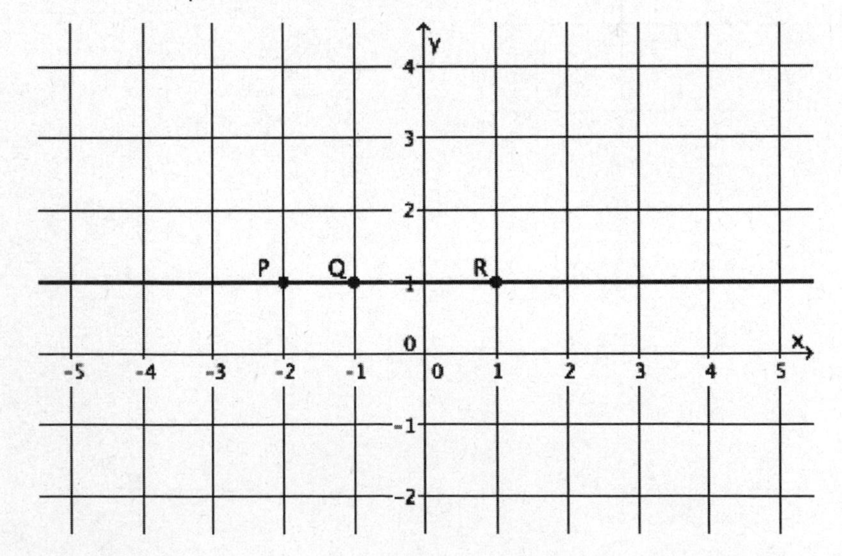

Lesson Summary

Slope is a number that can be used to describe the steepness of a line in a coordinate plane. The slope of a line is often represented by the symbol m.

Lines in a coordinate plane that are *left-to-right inclining* have a positive slope, as shown below.

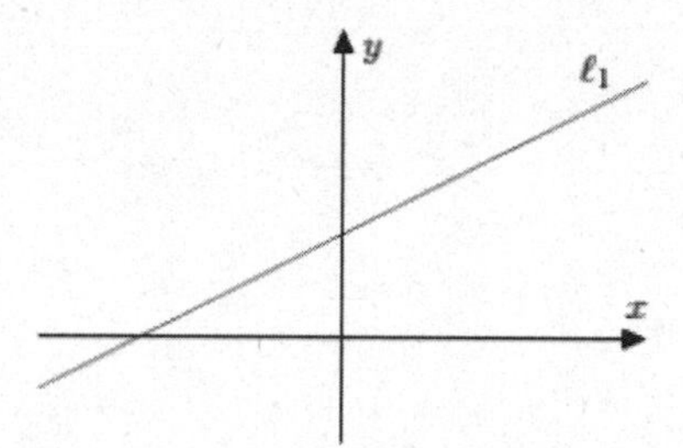

Lines in a coordinate plane that are *left-to-right declining* have a negative slope, as shown below.

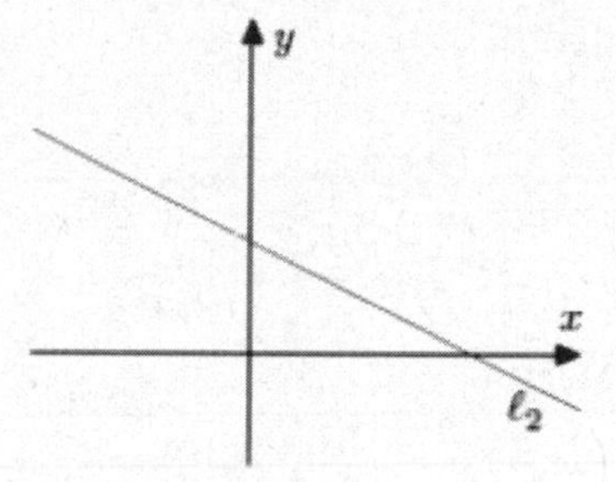

Determine the slope of a line when the horizontal distance between points is fixed at 1 by translating point Q to the origin of the graph and then identifying the y-coordinate of point R; by definition, that number is the slope of the line.

The slope of the line shown below is 2, so $m = 2$, because point R is at 2 on the y-axis.

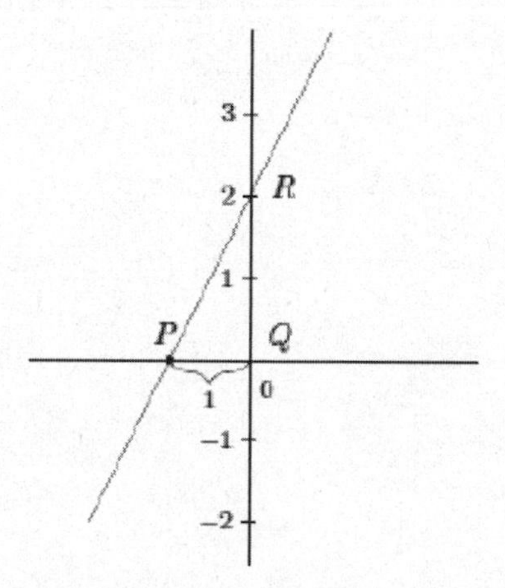

Name __________________________________ Date ________________

1. What is the slope of this non-vertical line? Use your transparency if needed.

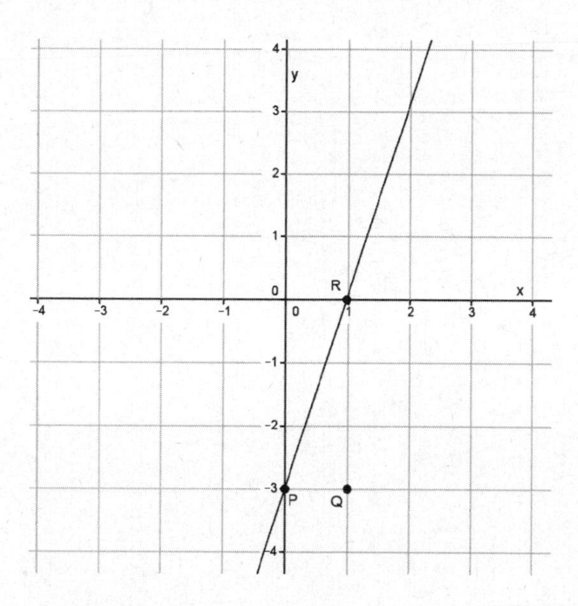

2. What is the slope of this non-vertical line? Use your transparency if needed.

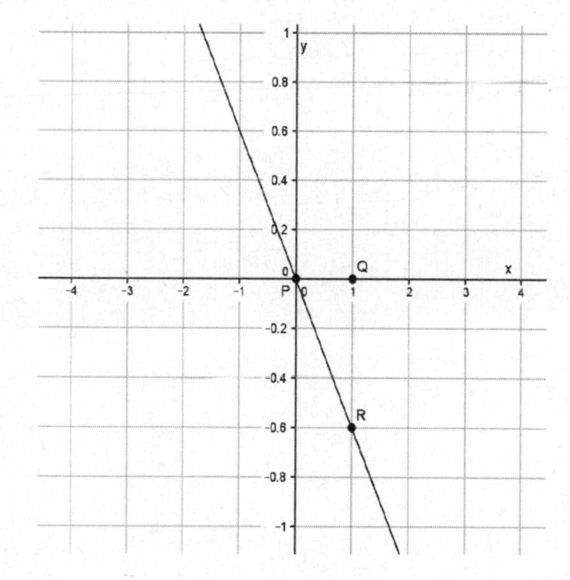

1. Does the graph of the line shown below have a positive or negative slope? Explain

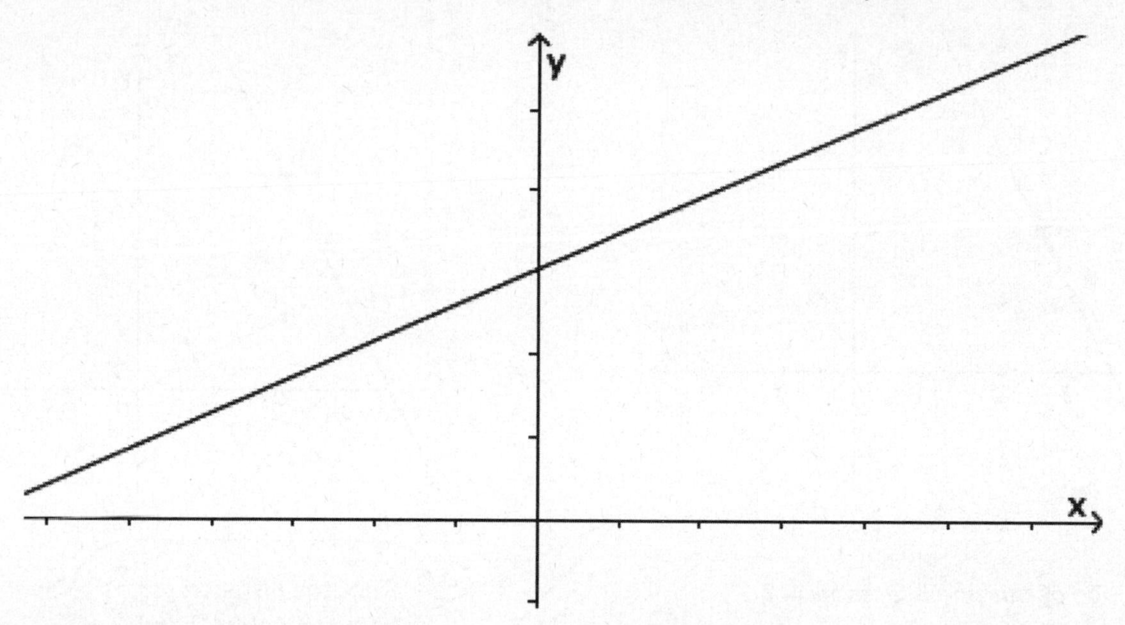

The graph of this line has a positive slope. It is left-to-right inclining, which is an indication of positive slope.

2. What is the slope of this non-vertical line? Use your transparency if needed.

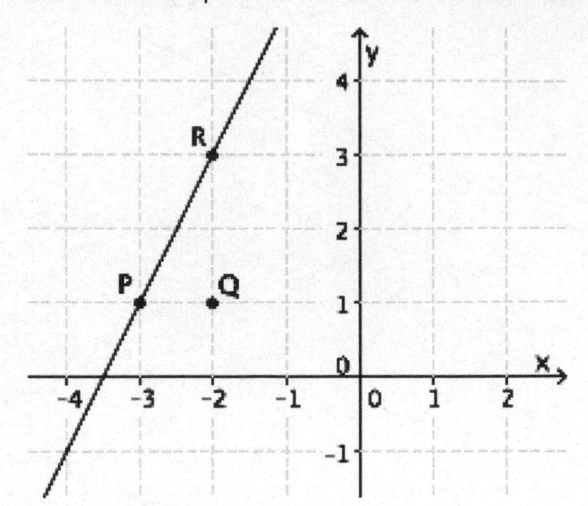

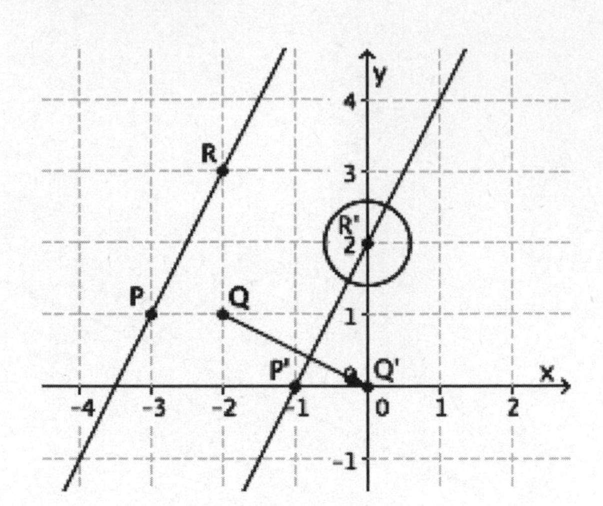

The slope of this line is 2, so $m = 2$.

Since the distance between points P and Q is 1 unit, I can trace everything onto a transparency and map point Q to the origin. The location of the translated point R gives me the slope of the line.

3. What is the slope of this non-vertical line? Use your transparency if needed.

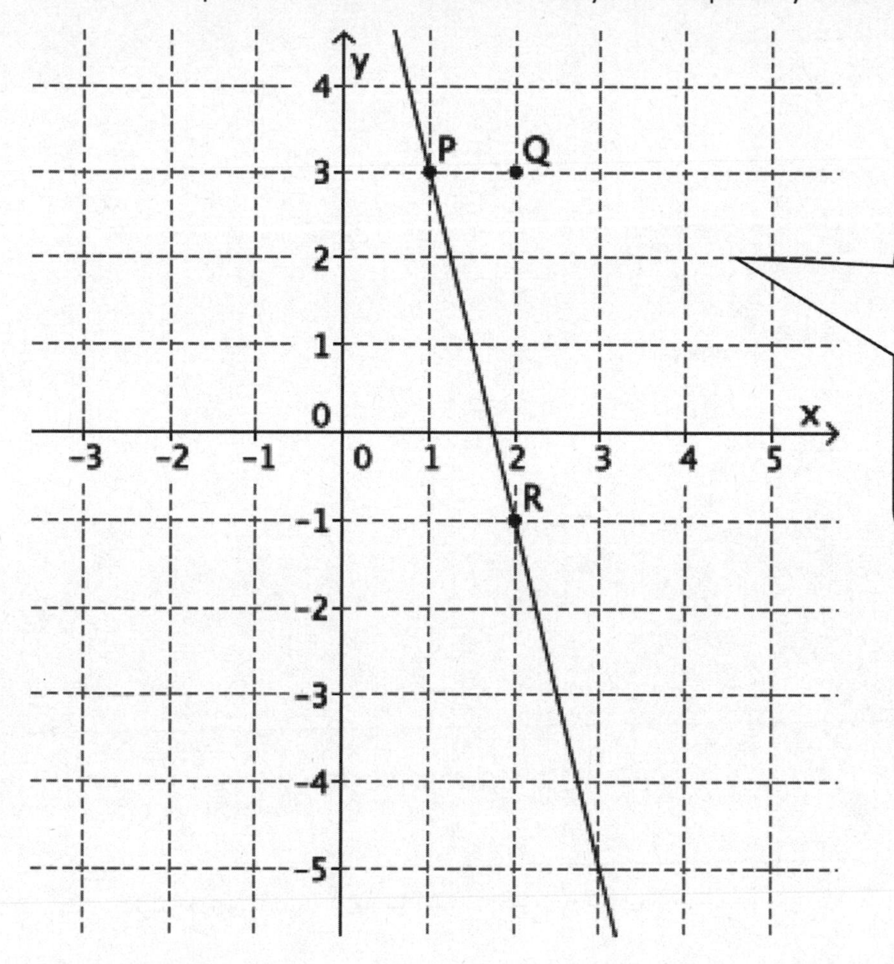

I can tell by the line that the slope will be negative. Just like I did in the last problem, I will use my transparency and translation to figure out the number that represents the slope.

The slope of this line is -4, so $m = -4$.

1. Does the graph of the line shown below have a positive or negative slope? Explain.

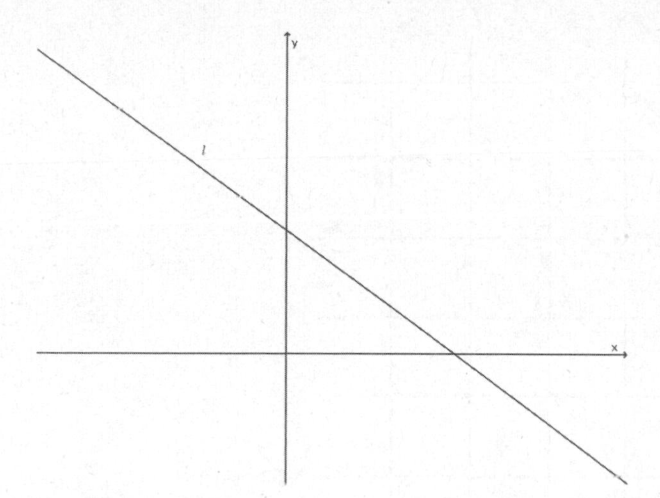

2. Does the graph of the line shown below have a positive or negative slope? Explain.

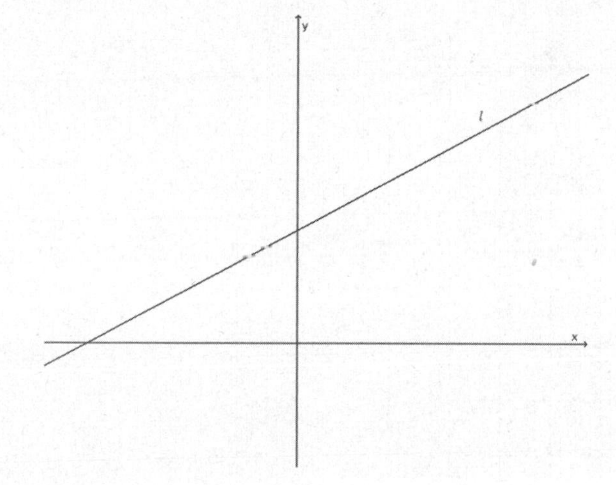

3. What is the slope of this non-vertical line? Use your transparency if needed.

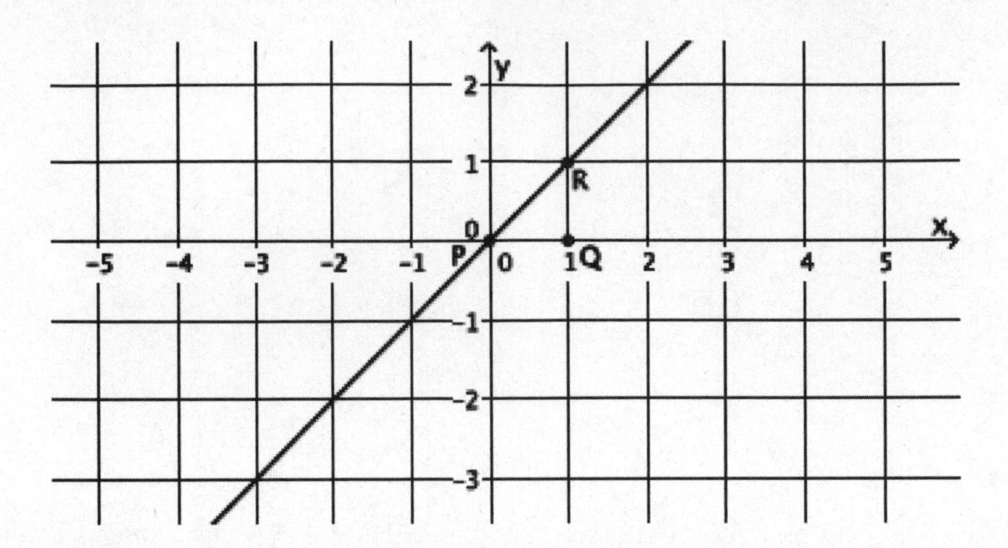

4. What is the slope of this non-vertical line? Use your transparency if needed.

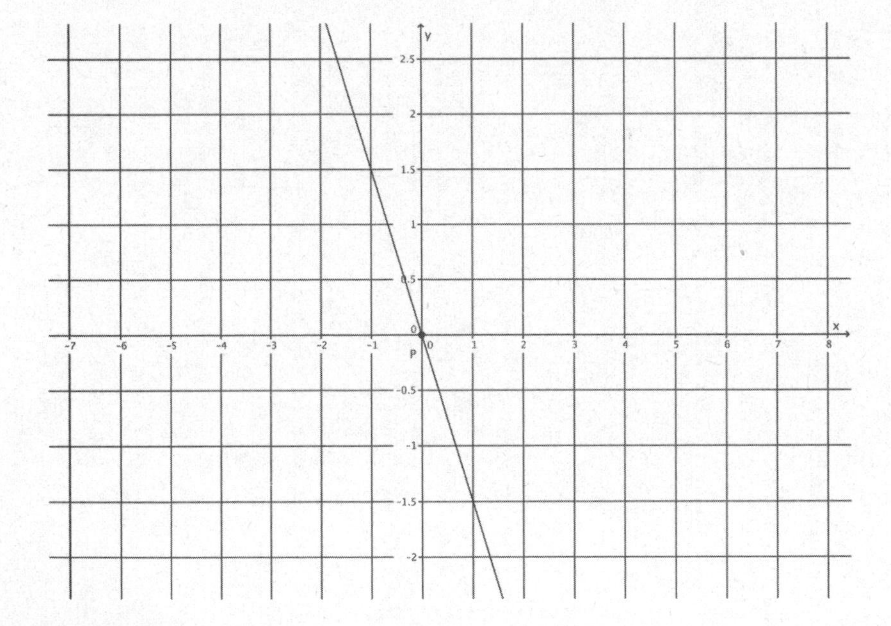

5. What is the slope of this non-vertical line? Use your transparency if needed.

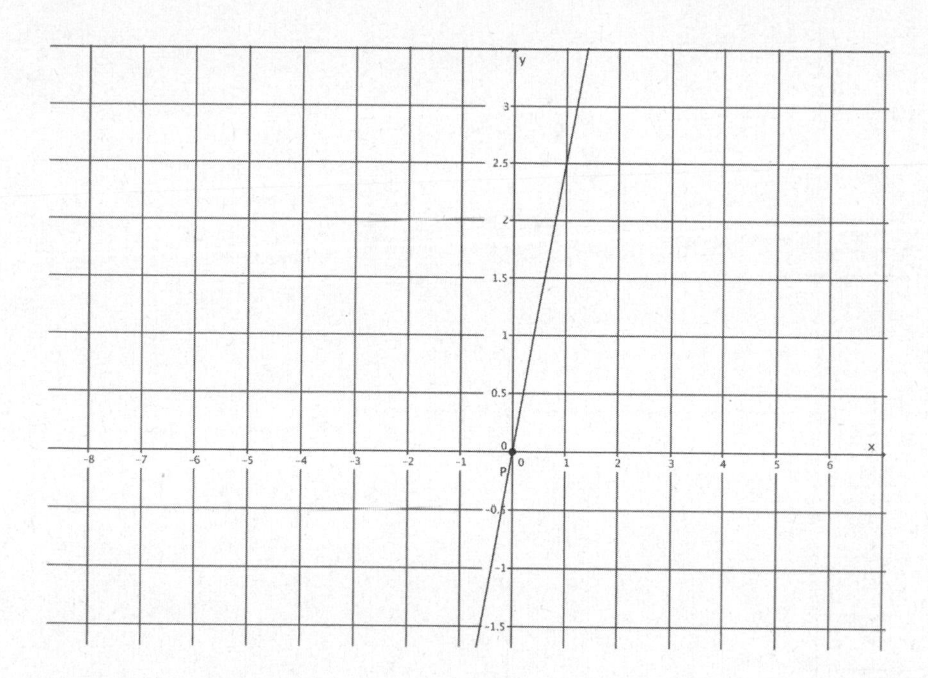

6. What is the slope of this non-vertical line? Use your transparency if needed.

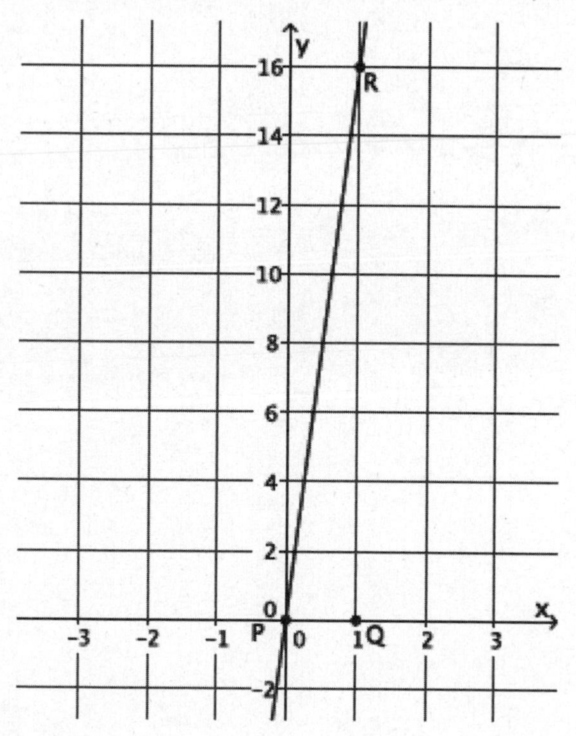

7. What is the slope of this non-vertical line? Use your transparency if needed.

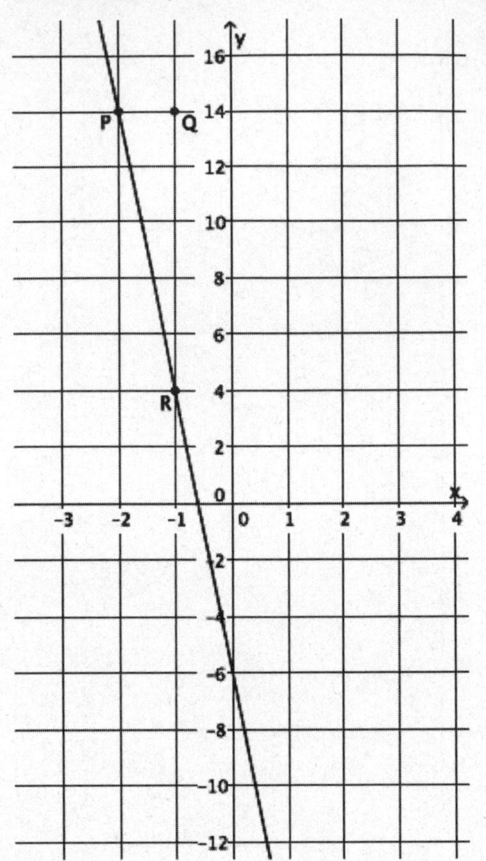

8. What is the slope of this non-vertical line? Use your transparency if needed.

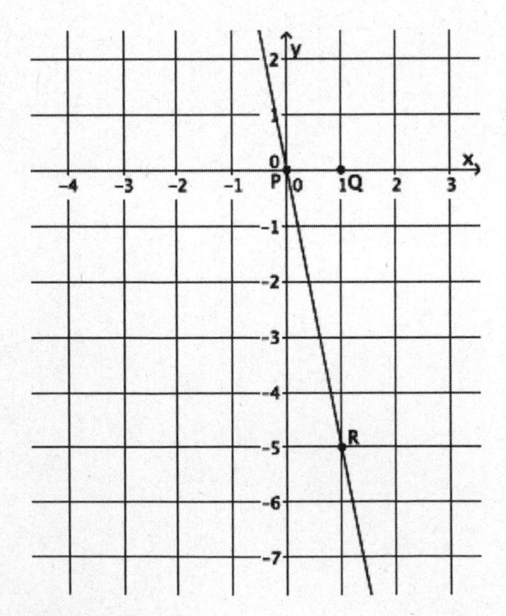

9. What is the slope of this non-vertical line? Use your transparency if needed.

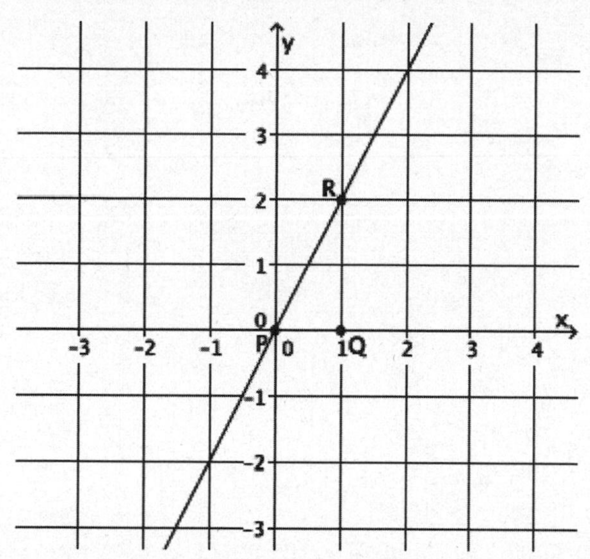

10. What is the slope of this non-vertical line? Use your transparency if needed.

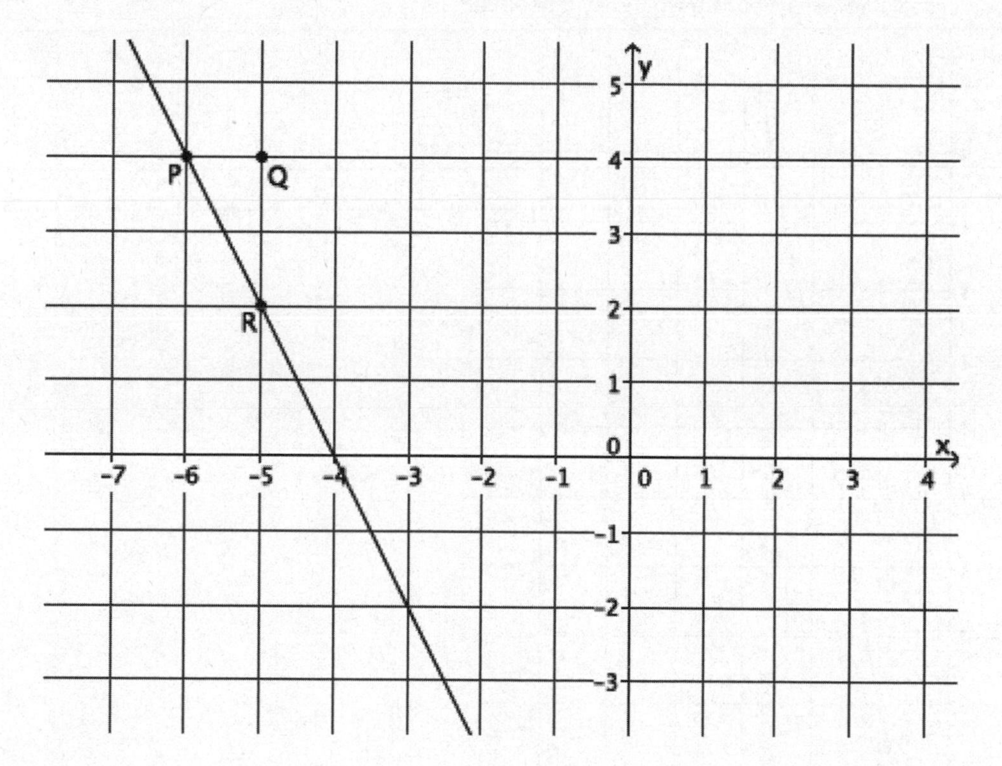

11. What is the slope of this non-vertical line? Use your transparency if needed.

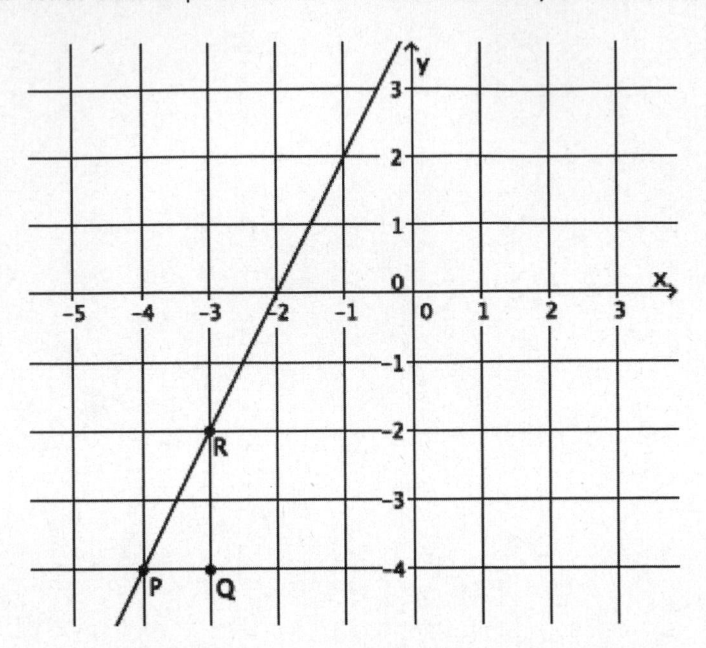

12. What is the slope of this non-vertical line? Use your transparency if needed.

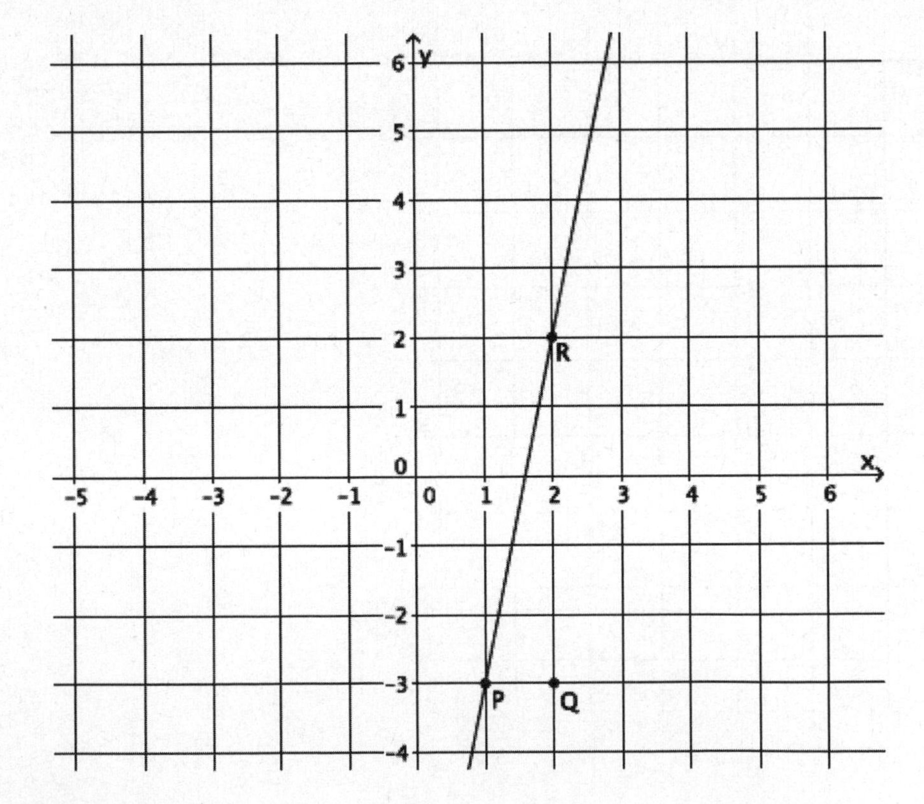

13. What is the slope of this non-vertical line? Use your transparency if needed.

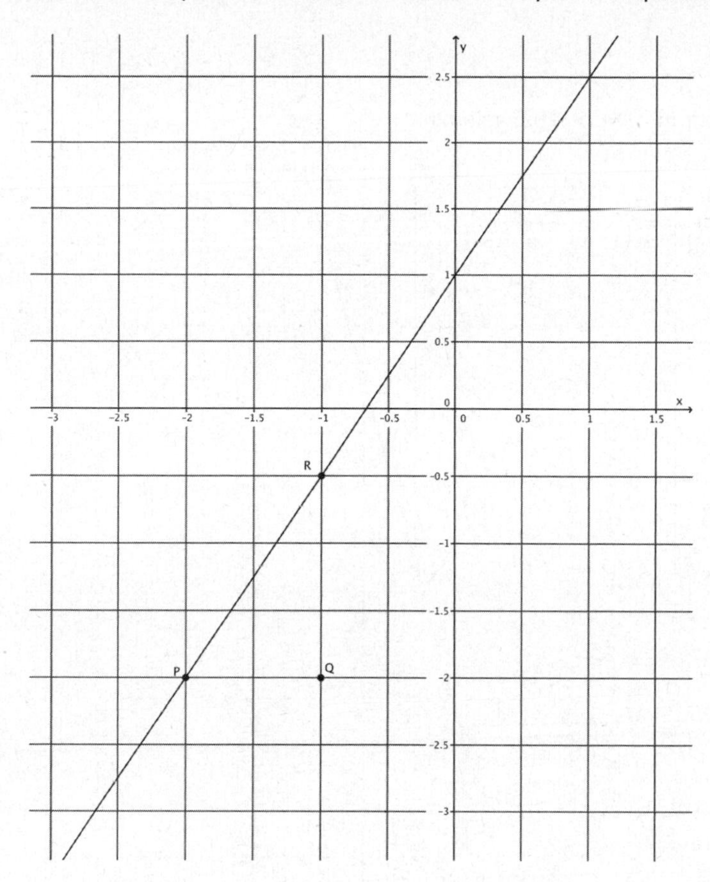

14. What is the slope of this non-vertical line? Use your transparency if needed.

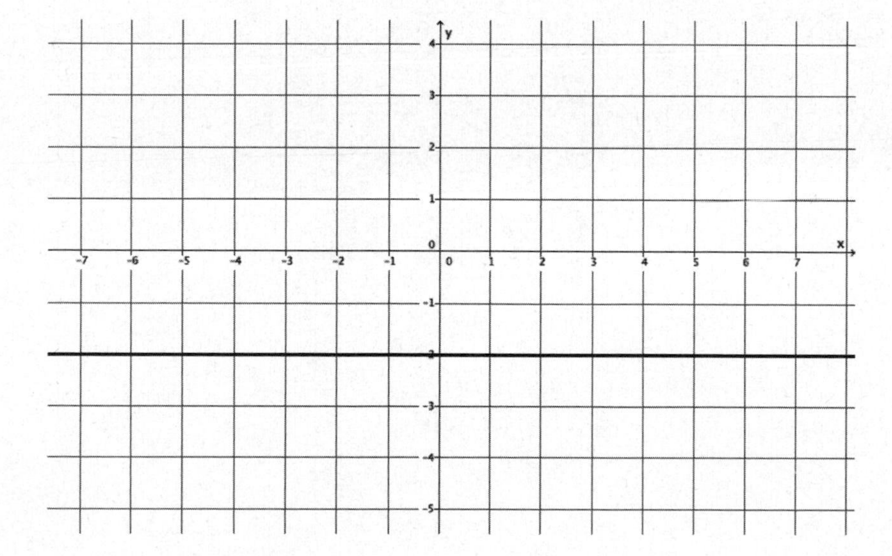

In Lesson 11, you did the work below involving constant rate problems. Use the table and the graphs provided to answer the questions that follow.

15. Suppose the volume of water that comes out in three minutes is 10.5 gallons.

t (time in minutes)	**Linear Equation: $V = \frac{10.5}{3}t$**	**V (in gallons)**
0	$V = \frac{10.5}{3}(0)$	0
1	$V = \frac{10.5}{3}(1)$	$\frac{10.5}{3} = 3.5$
2	$V = \frac{10.5}{3}(2)$	$\frac{21}{3} = 7$
3	$V = \frac{10.5}{3}(3)$	$\frac{31.5}{3} = 10.5$
4	$V = \frac{10.5}{3}(4)$	$\frac{42}{3} = 14$

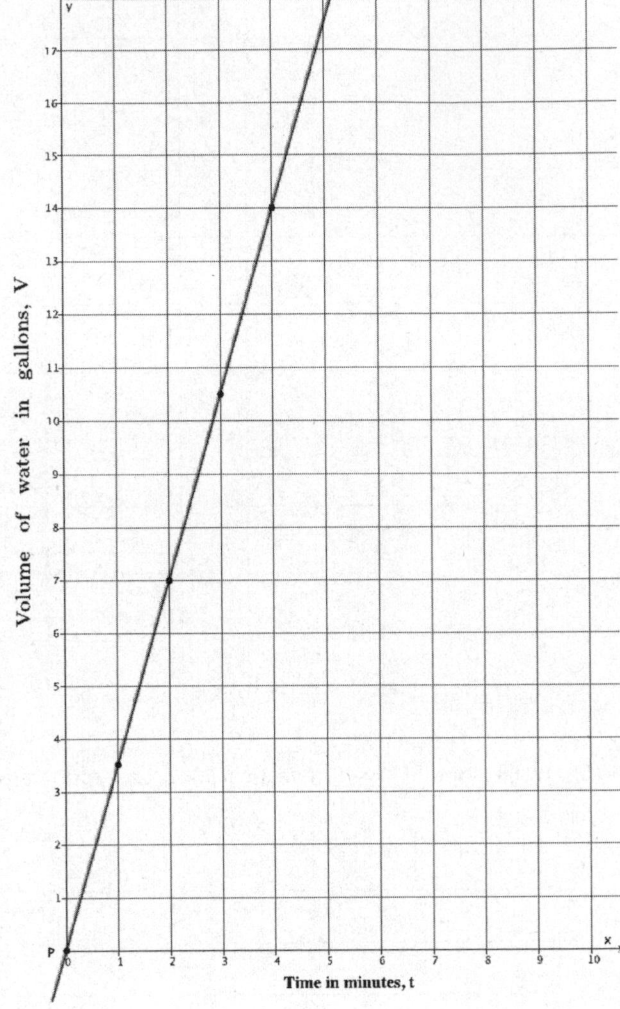

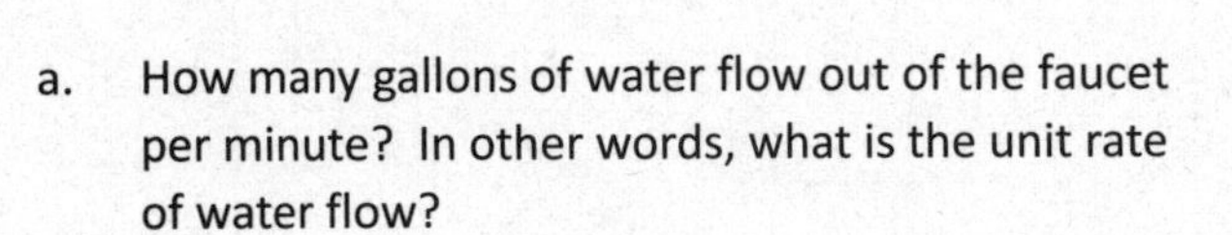

a. How many gallons of water flow out of the faucet per minute? In other words, what is the unit rate of water flow?

b. Assume that the graph of the situation is a line, as shown in the graph. What is the slope of the line?

16. Emily paints at a constant rate. She can paint 32 square feet in five minutes.

t (time in minutes)	Linear Equation: $A = \frac{32}{5}t$	A (area painted in square feet
0	$A = \frac{32}{5}(0)$	0
1	$A = \frac{32}{5}(1)$	$\frac{32}{5} = 6.4$
2	$A = \frac{32}{5}(2)$	$\frac{64}{5} = 12.8$
3	$A = \frac{32}{5}(3)$	$\frac{96}{5} = 19.2$
4	$A = \frac{32}{5}(4)$	$\frac{128}{5} = 25.6$

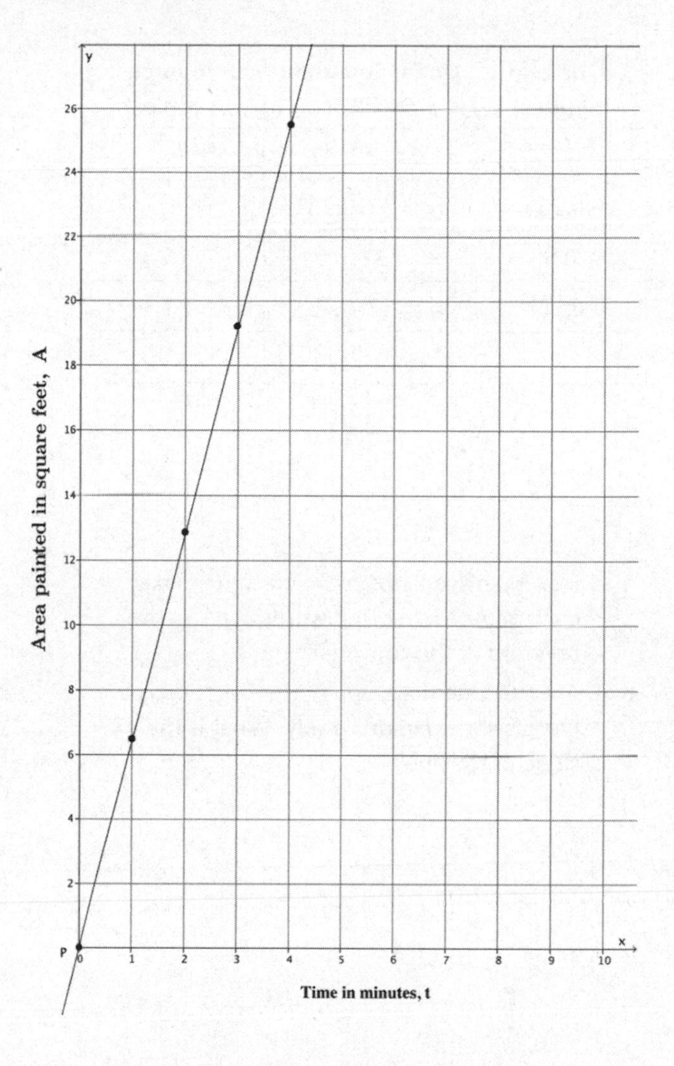

a. How many square feet can Emily paint in one minute? In other words, what is her unit rate of painting?

b. Assume that the graph of the situation is a line, as shown in the graph. What is the slope of the line?

17. A copy machine makes copies at a constant rate. The machine can make 80 copies in $2\frac{1}{2}$ minutes.

t (time in minutes)	Linear Equation: $n = 32t$	n (number of copies)
0	$n = 32(0)$	0
0.25	$n = 32(0.25)$	8
0.5	$n = 32(0.5)$	16
0.75	$n = 32(0.75)$	24
1	$n = 32(1)$	32

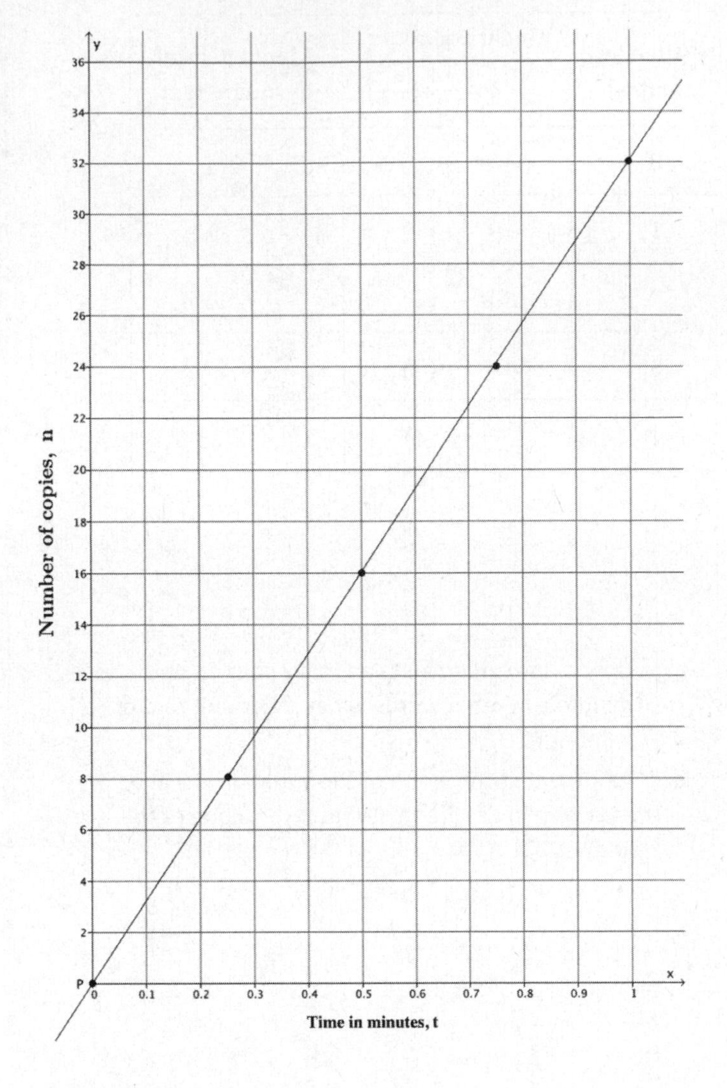

a. How many copies can the machine make each minute? In other words, what is the unit rate of the copy machine?

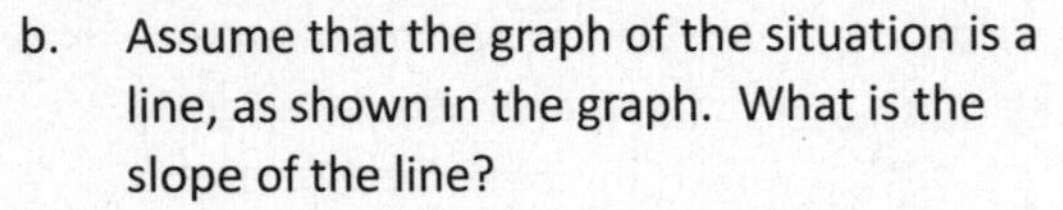

b. Assume that the graph of the situation is a line, as shown in the graph. What is the slope of the line?

Example 1

Using what you learned in the last lesson, determine the slope of the line with the following graph.

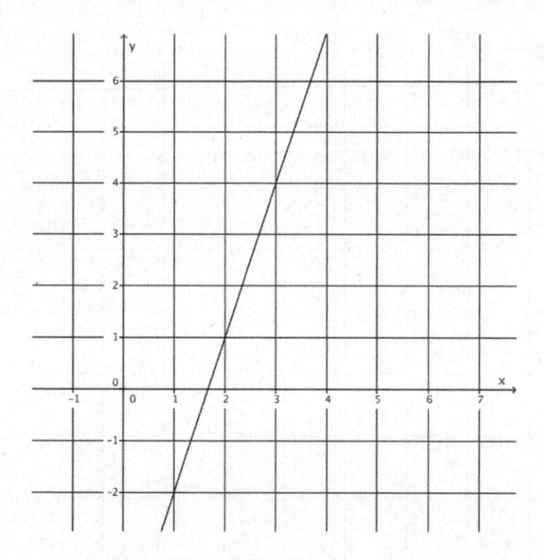

Example 2

Using what you learned in the last lesson, determine the slope of the line with the following graph.

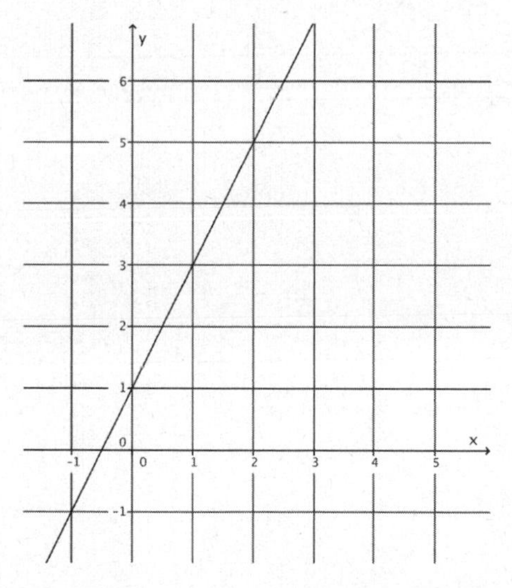

Example 3

What is different about this line compared to the last two examples?

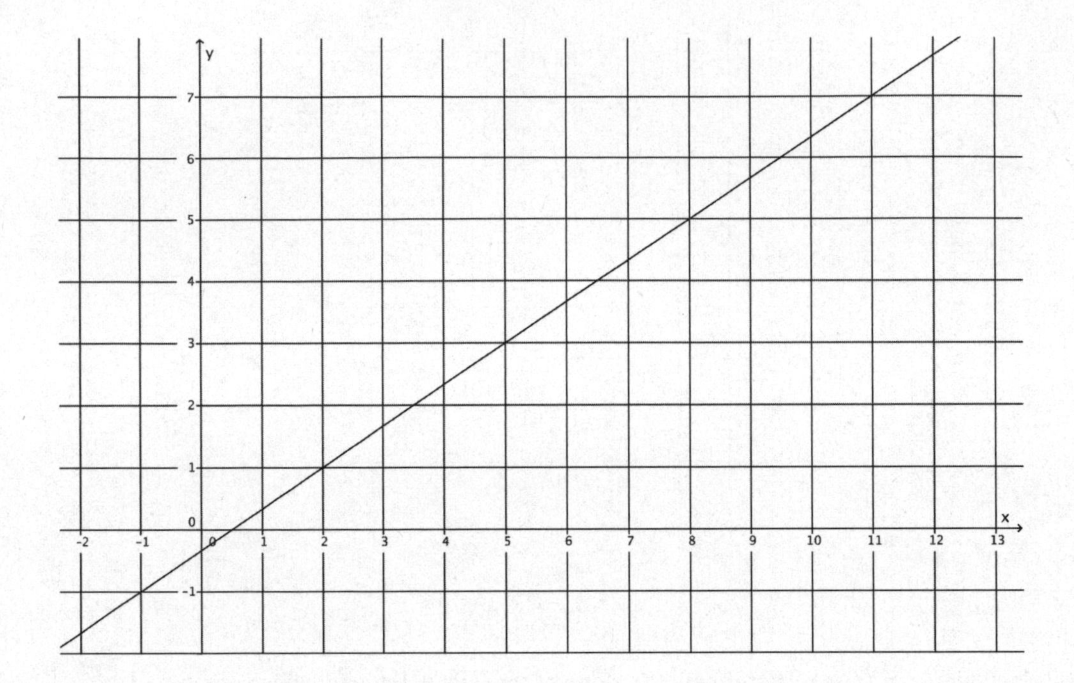

Exercise

Let's investigate concretely to see if the claim that we can find slope between any two points is true.

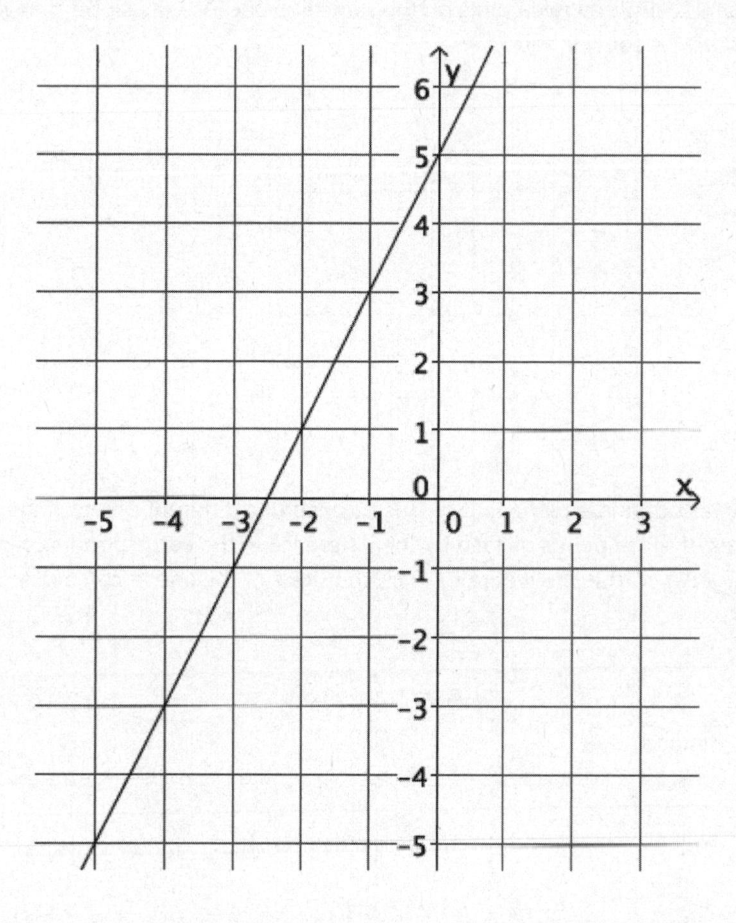

a. Select any two points on the line to label as P and R.

b. Identify the coordinates of points P and R.

c. Find the slope of the line using as many different points as you can. Identify your points, and show your work below.

Lesson Summary

The slope of a line can be calculated using *any* two points on the same line because the slope triangles formed are similar, and corresponding sides will be equal in ratio.

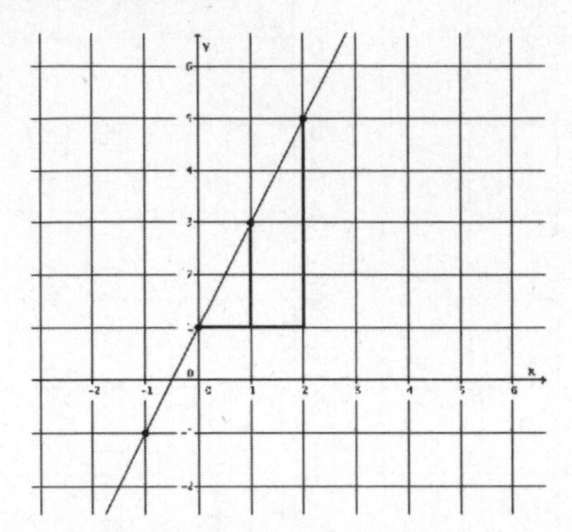

The *slope* of a non-vertical line in a coordinate plane that passes through two different points is the number given by the difference in y-coordinates of those points divided by the difference in the corresponding x-coordinates. For two points $P(p_1, p_2)$ and $R(r_1, r_2)$ on the line where $p_1 \neq r_1$, the slope of the line m can be computed by the formula

$$m = \frac{p_2 - r_2}{p_1 - r_1}.$$

The slope of a vertical line is not defined.

Name ____ Date ____________________

Find the rate of change of the line by completing parts (a) and (b).

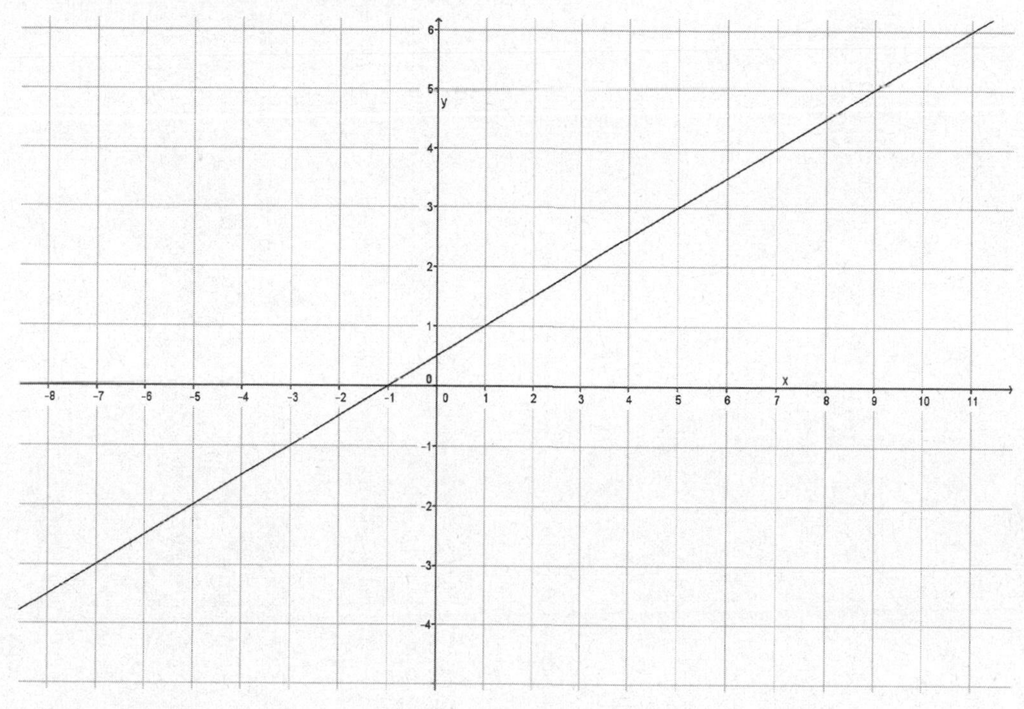

a. Select any two points on the line to label as P and R. Name their coordinates.

b. Compute the rate of change of the line.

1. Calculate the slope of the line using two different pairs of points.

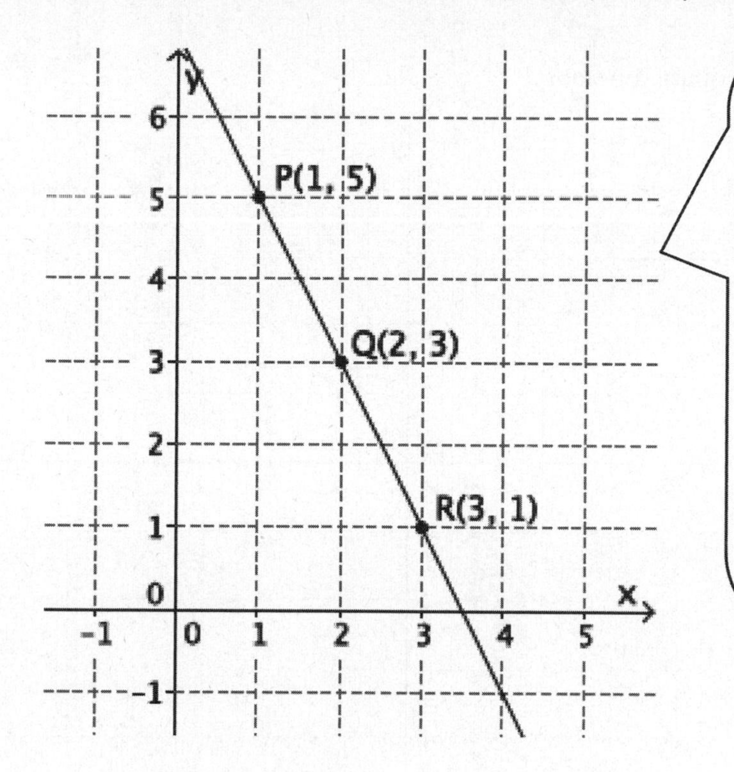

I need to choose three points on the line. The points $P(p_1, p_2)$ and $Q(q_1, q_2)$ are used in my first slope equation. I need to remember that no matter how the slope is written, the difference in the 2nd values (y-values) is in the numerator of the slope, and the difference in the 1st values (x-values) is in the denominator of the slope. The equation $m = \frac{q_2 - p_2}{p_1 - q_1}$ would be wrong since the values of point Q do not come first in each difference.

$$m = \frac{p_2 - q_2}{p_1 - q_1} = \frac{5-3}{1-2} = \frac{2}{-1} = -2$$

$$m = \frac{r_2 - q_2}{r_1 - q_1} = \frac{1-3}{3-2} = \frac{-2}{1} = -2$$

2. Calculate the slope of the line using two different pairs of points.

   a. Select any two points on the line to compute the slope.

   Let the two points be $P(-3, -3)$ and $Q(-1, -1)$.

   $$m = \frac{p_2 - q_2}{p_1 - q_1}$$
   $$= \frac{-3 - (-1)}{-3 - (-1)}$$
   $$= \frac{-2}{-2}$$
   $$= 1$$

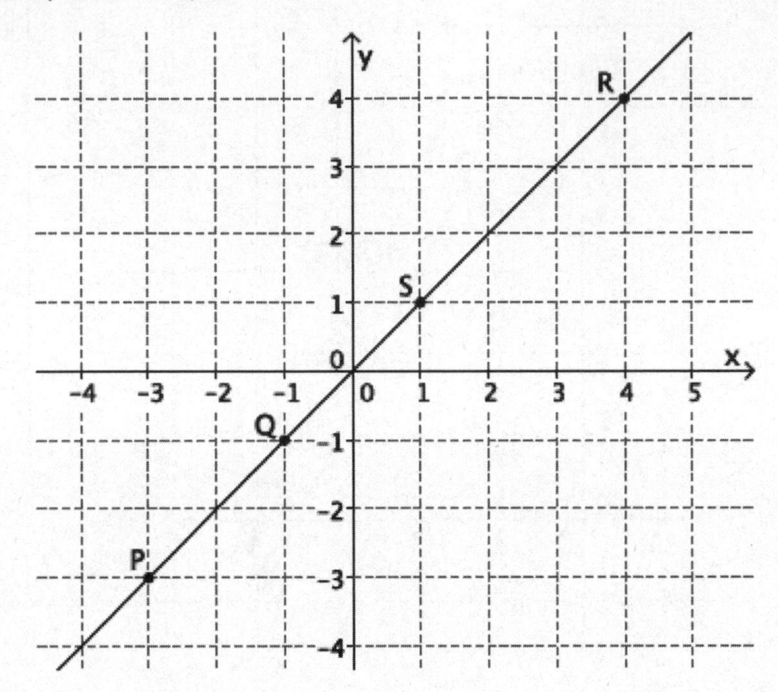

   b. Select two different points on the line to calculate the slope.

   Let the two points be $S(1, 1)$ and $R(4, 4)$.

   $$m = \frac{s_2 - r_2}{s_1 - r_1}$$
   $$= \frac{1 - 4}{1 - 4}$$
   $$= \frac{-3}{-3}$$
   $$= 1$$

   c. What do you notice about your answers in parts (a) and (b)? Explain.

   The slopes are equal in parts (a) and (b). This is true because of what we know about similar triangles. The slope triangle that is drawn between the two points selected in part (a) is similar to the slope triangle that is drawn between the two points in part (b) by the AA criterion. Then, because the corresponding sides of similar triangles are equal in ratio, the slopes are equal.

3. Your teacher tells you that a line goes through the points $\left(1, \frac{3}{4}\right)$ and $(-2, -3)$.

   a. Calculate the slope of this line.

$$m = \frac{p_2 - r_2}{p_1 - r_1}$$
$$= \frac{\frac{3}{4} - (-3)}{1 - (-2)}$$
$$= \frac{3\frac{3}{4}}{3}$$
$$= \frac{5}{4}$$

   b. Do you think the slope will be the same if the order of the points is reversed? Verify by calculating the slope, and explain your result.

   The slope should be the same because we are joining the same two points. Since the slope of a line can be computed using any two points on the same line, it makes sense that it does not matter which point we name as P and which point we name as R.

$$m = \frac{r_2 - p_2}{r_1 - p_1}$$
$$= \frac{-3 - \frac{3}{4}}{-2 - 1}$$
$$= \frac{-3\frac{3}{4}}{-3}$$
$$= \frac{5}{4}$$

4. Each of the lines in the lesson was non-vertical. Consider the slope of a vertical line, $x = -2$. Select two points on the line to calculate slope. Based on your answer, why do you think the topic of slope focuses only on non-vertical lines?

$$m = \frac{r_2 - p_2}{r_1 - p_1}$$

$$= \frac{-1 - 5}{-2 - (-2)}$$

$$= \frac{-6}{0}$$

The computation of slope using the formula leads to a fraction with zero as its denominator, which is undefined. The topic of slope does not focus on vertical lines because the slope of a vertical line is undefined.

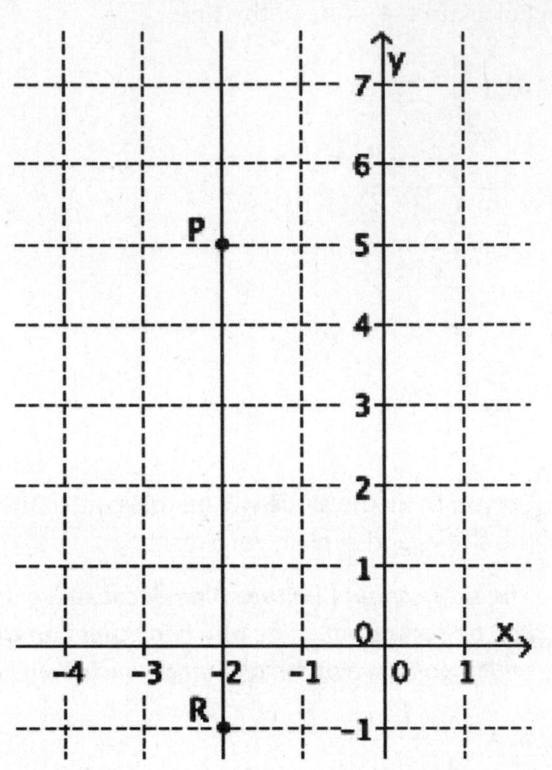

1. Calculate the slope of the line using two different pairs of points.

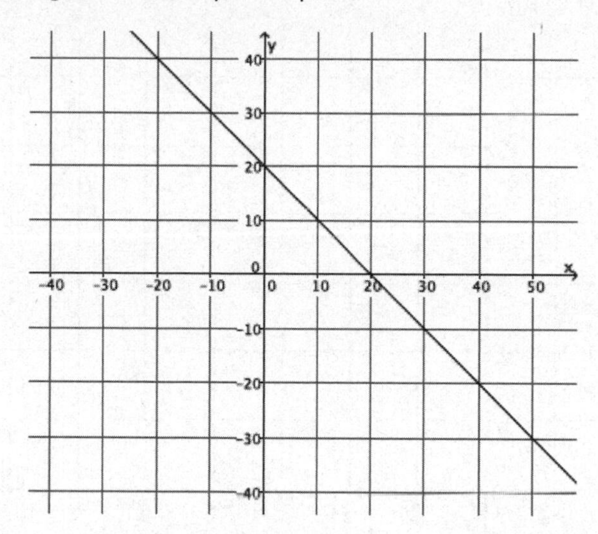

2. Calculate the slope of the line using two different pairs of points.

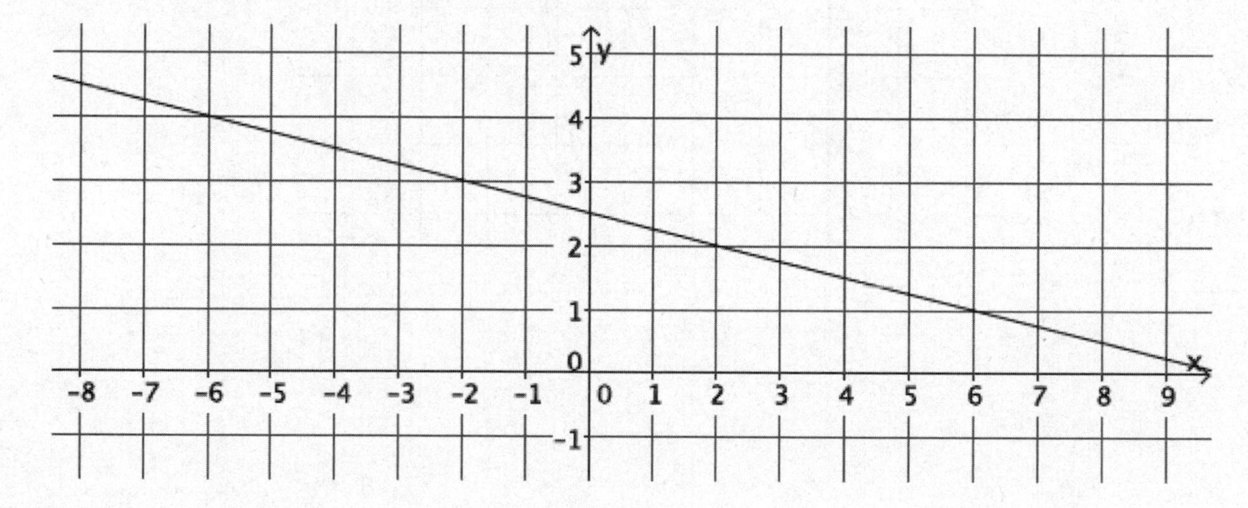

3. Calculate the slope of the line using two different pairs of points.

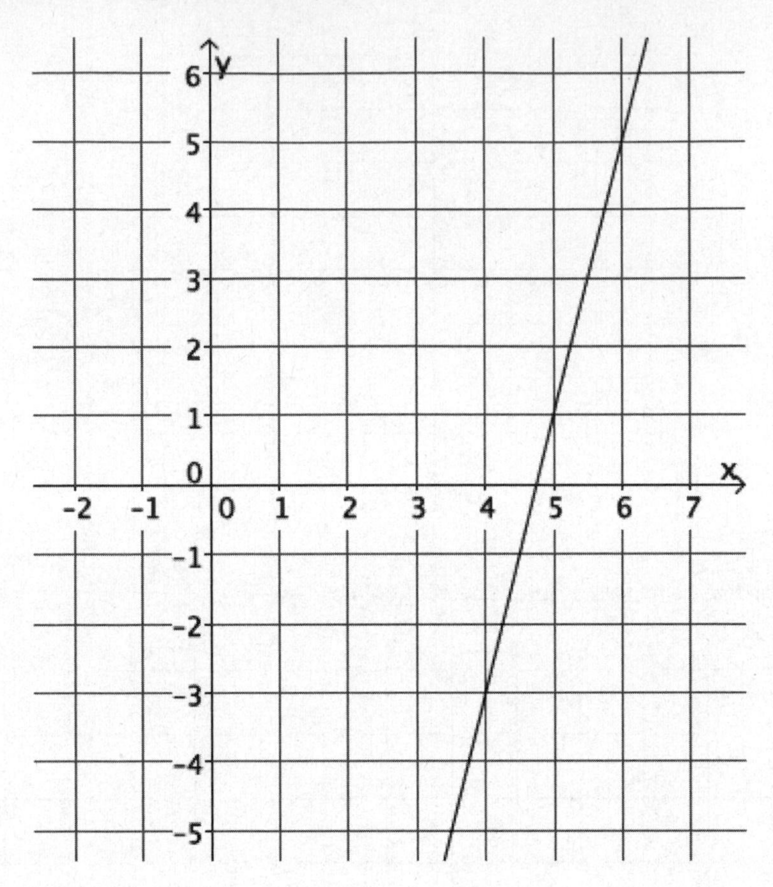

4. Calculate the slope of the line using two different pairs of points.

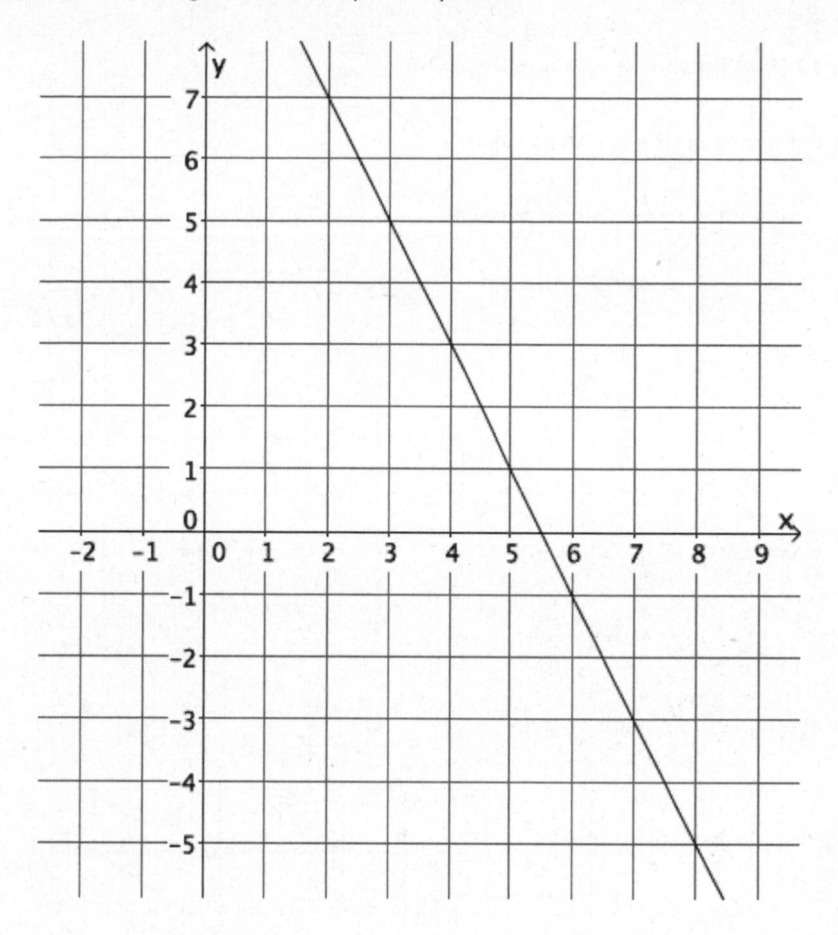

5. Calculate the slope of the line using two different pairs of points.

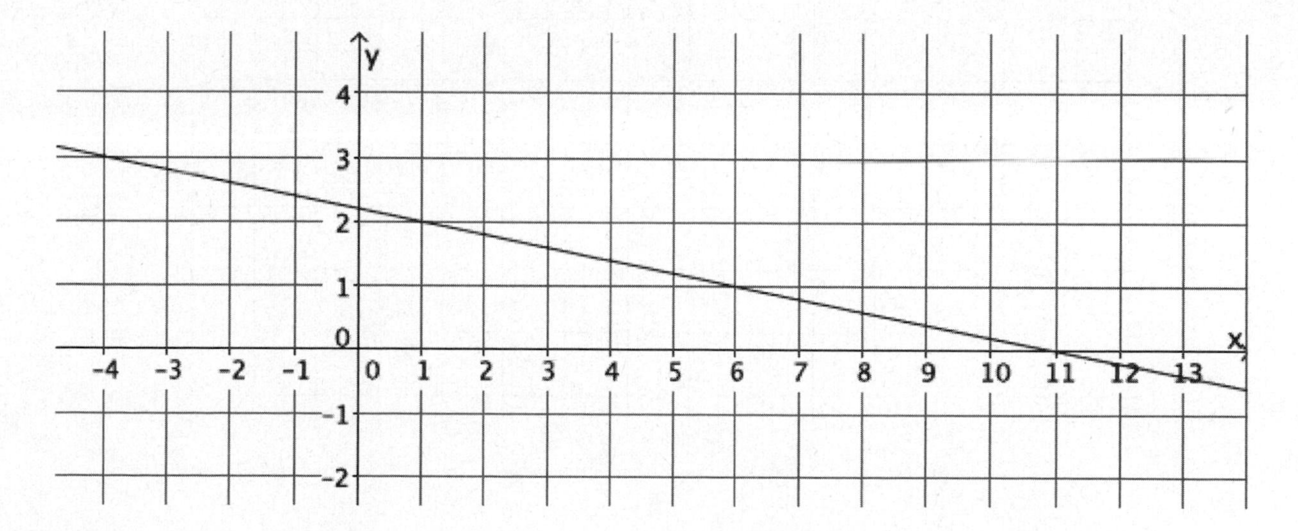

6. Calculate the slope of the line using two different pairs of points.
   a. Select any two points on the line to compute the slope.
   b. Select two different points on the line to calculate the slope.
   c. What do you notice about your answers in parts (a) and (b)? Explain.

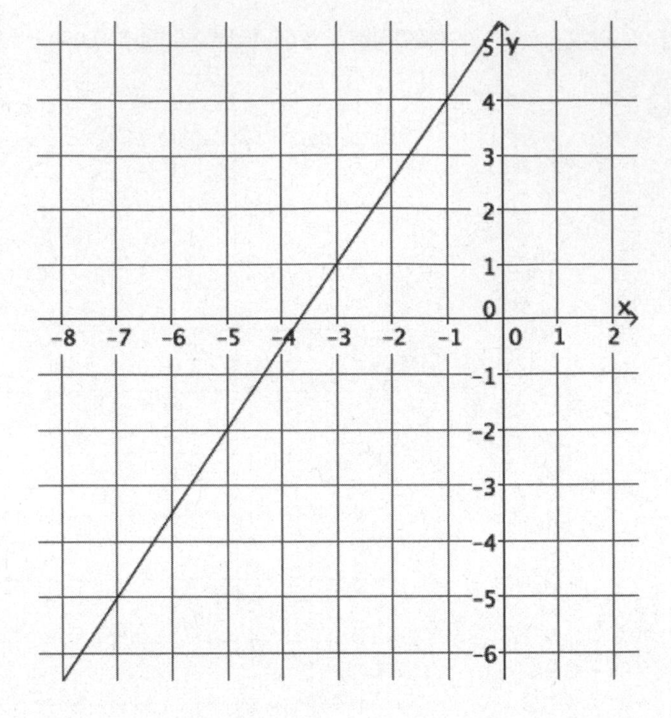

7. Calculate the slope of the line in the graph below.

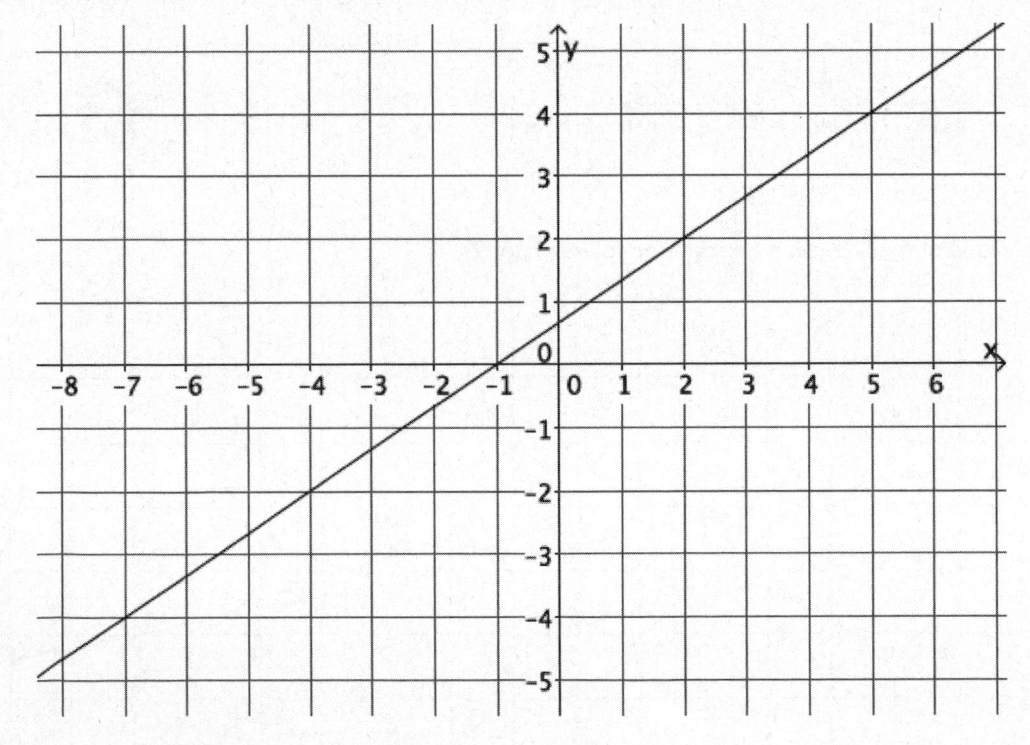

8. Your teacher tells you that a line goes through the points $\left(-6, \frac{1}{2}\right)$ and $(-4, 3)$.
   a. Calculate the slope of this line.
   b. Do you think the slope will be the same if the order of the points is reversed? Verify by calculating the slope, and explain your result.

9. Use the graph to complete parts (a)–(c).
   a. Select any two points on the line to calculate the slope.
   b. Compute the slope again, this time reversing the order of the coordinates.
   c. What do you notice about the slopes you computed in parts (a) and (b)?
   d. Why do you think $m = \frac{(p_2 - r_2)}{(p_1 - r_1)} = \frac{(r_2 - p_2)}{(r_1 - p_1)}$?

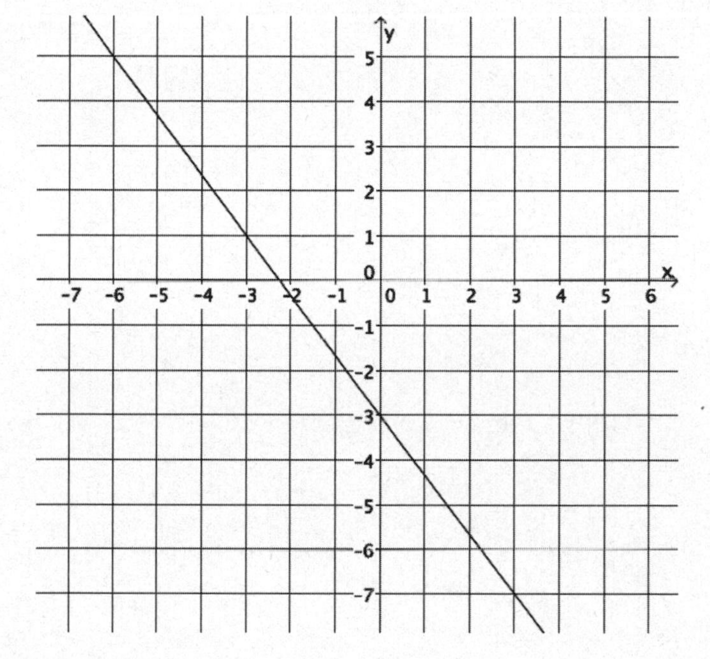

10. Each of the lines in the lesson was non-vertical. Consider the slope of a vertical line, $x = 2$. Select two points on the line to calculate slope. Based on your answer, why do you think the topic of slope focuses only on non-vertical lines?

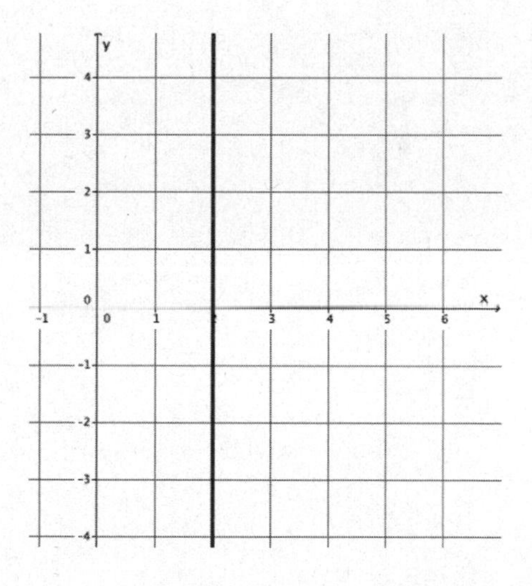

Challenge:

11. A certain line has a slope of $\frac{1}{2}$. Name two points that may be on the line.

Exercises

1. Find at least three solutions to the equation $y = 2x$, and graph the solutions as points on the coordinate plane. Connect the points to make a line. Find the slope of the line.

2. Find at least three solutions to the equation $y = 3x - 1$, and graph the solutions as points on the coordinate plane. Connect the points to make a line. Find the slope of the line.

3. Find at least three solutions to the equation $y = 3x + 1$, and graph the solutions as points on the coordinate plane. Connect the points to make a line. Find the slope of the line.

4. The graph of the equation $y = 7x - 3$ has what slope?

5. The graph of the equation $y = -\frac{3}{4}x - 3$ has what slope?

6. You have $20 in savings at the bank. Each week, you add $2 to your savings. Let y represent the total amount of money you have saved at the end of x weeks. Write an equation to represent this situation, and identify the slope of the equation. What does that number represent?

7. A friend is training for a marathon. She can run 4 miles in 28 minutes. Assume she runs at a constant rate. Write an equation to represent the total distance, y, your friend can run in x minutes. Identify the slope of the equation. What does that number represent?

8. Four boxes of pencils cost $\$5$. Write an equation that represents the total cost, y, for x boxes of pencils. What is the slope of the equation? What does that number represent?

9. Solve the following equation for y, and then identify the slope of the line: $9x - 3y = 15$.

10. Solve the following equation for y, and then identify the slope of the line: $5x + 9y = 8$.

11. Solve the following equation for y, and then identify the slope of the line: $ax + by = c$.

Lesson Summary

The line joining two distinct points of the graph of the linear equation $y = mx + b$ has slope m.

The m of $y = mx + b$ is the number that describes the slope. For example, in the equation $y = -2x + 4$, the slope of the graph of the line is -2.

Name ____________________ Date ____________

1. Solve the following equation for y: $35x - 7y = 49$.

2. What is the slope of the equation in Problem 1?

3. Show, using similar triangles, why the graph of an equation of the form $y = mx$ is a line with slope m.

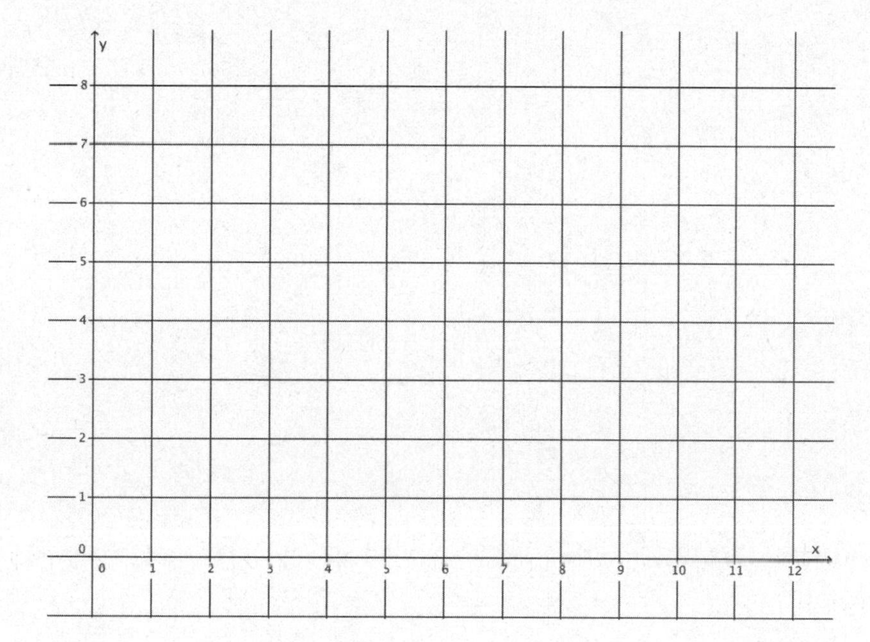

1. Solve the following equation for y: $-3x + 9y = 18$. Then, answer the questions that follow.

$$-3x + 9y = 18$$
$$-3x + 3x + 9y = 18 + 3x$$
$$9y = 18 + 3x$$
$$\frac{9}{9}y = \frac{18}{9} + \frac{3}{9}x$$
$$y = 2 + \frac{1}{3}x$$
$$y = \frac{1}{3}x + 2$$

a. Based on your transformed equation, what is the slope of the linear equation $-3x + 9y = 18$?

The slope is $\frac{1}{3}$.

b. Complete the table to find solutions to the linear equation.

Since the slope is a fraction, $\frac{1}{3}$, I need to choose x-values that are multiples of 3.

x	Transformed Equation: $y = \frac{1}{3}x + 2$	y
-3	$y = \frac{1}{3}(-3) + 2$ $= -1 + 2$ $= 1$	1
0	$y = \frac{1}{3}(0) + 2$ $= 2$	2
3	$y = \frac{1}{3}(3) + 2$ $= 1 + 2$ $= 3$	3
6	$y = \frac{1}{3}(6) + 2$ $= 2 + 2$ $= 4$	4

c. Graph the points on the coordinate plane.

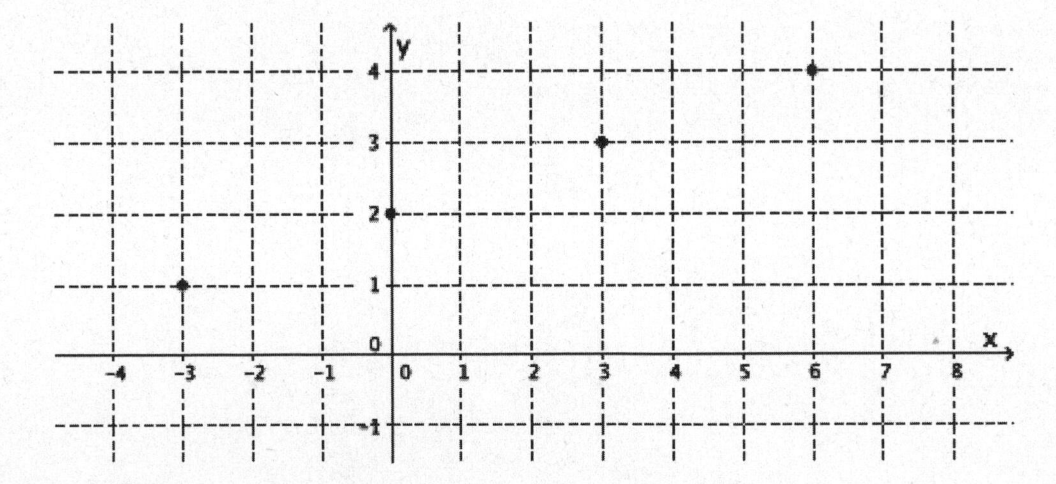

d. Find the slope between any two points.

Using the points $(-3, 1)$ ***and*** $(3, 3)$,

$$m = \frac{1-3}{-3-3}$$
$$= \frac{-2}{-6}$$
$$= \frac{1}{3}$$

e. The slope you found in part (d) should be equal to the slope you noted in part (a). If so, connect the points to make the line that is the graph of an equation of the form $y = mx + b$ that has slope m.

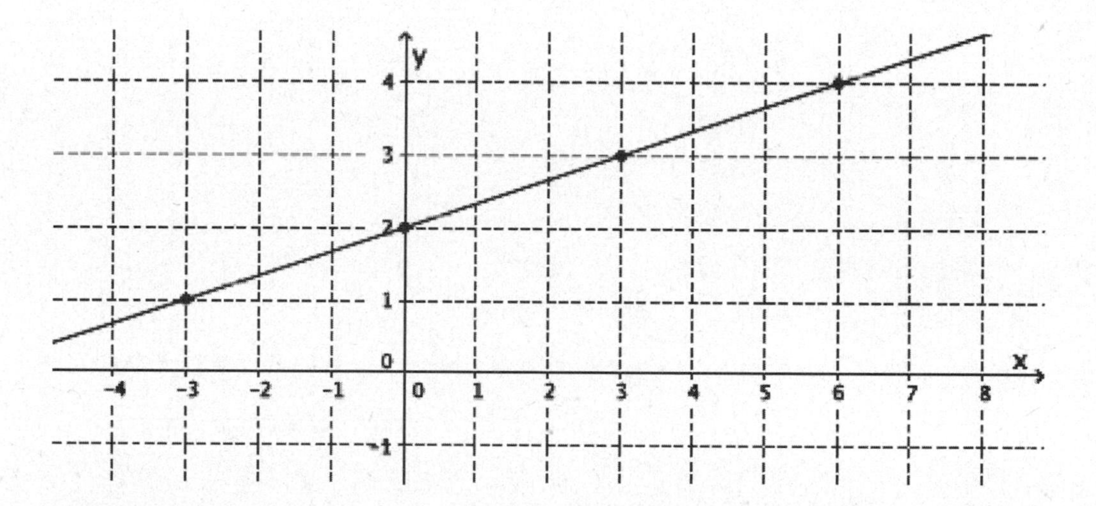

f. Note the location (ordered pair) that describes where the line intersects the y-axis.

$(0, 2)$ ***is the location where the line intersects the y-axis.***

1. Solve the following equation for y: $-4x + 8y = 24$. Then, answer the questions that follow.

   a. Based on your transformed equation, what is the slope of the linear equation $-4x + 8y = 24$?

   b. Complete the table to find solutions to the linear equation.

x	Transformed Linear Equation:	y

   c. Graph the points on the coordinate plane.

   d. Find the slope between any two points.

   e. The slope you found in part (d) should be equal to the slope you noted in part (a). If so, connect the points to make the line that is the graph of an equation of the form $y = mx + b$ that has slope m.

   f. Note the location (ordered pair) that describes where the line intersects the y-axis.

2. Solve the following equation for y: $9x + 3y = 21$. Then, answer the questions that follow.

   a. Based on your transformed equation, what is the slope of the linear equation $9x + 3y = 21$?

   b. Complete the table to find solutions to the linear equation.

x	Transformed Linear Equation:	y

   c. Graph the points on the coordinate plane.

   d. Find the slope between any two points.

   e. The slope you found in part (d) should be equal to the slope you noted in part (a). If so, connect the points to make the line that is the graph of an equation of the form $y = mx + b$ that has slope m.

   f. Note the location (ordered pair) that describes where the line intersects the y-axis.

3. Solve the following equation for y: $2x + 3y = -6$. Then, answer the questions that follow.

   a. Based on your transformed equation, what is the slope of the linear equation $2x + 3y = -6$?

   b. Complete the table to find solutions to the linear equation.

x	**Transformed Linear Equation:**	y

   c. Graph the points on the coordinate plane.

   d. Find the slope between any two points.

   e. The slope you found in part (d) should be equal to the slope you noted in part (a). If so, connect the points to make the line that is the graph of an equation of the form $y = mx + b$ that has slope m.

   f. Note the location (ordered pair) that describes where the line intersects the y-axis.

4. Solve the following equation for y: $5x - y = 4$. Then, answer the questions that follow.

   a. Based on your transformed equation, what is the slope of the linear equation $5x - y = 4$?

   b. Complete the table to find solutions to the linear equation.

x	**Transformed Linear Equation:**	y

   c. Graph the points on the coordinate plane.

   d. Find the slope between any two points.

   e. The slope you found in part (d) should be equal to the slope you noted in part (a). If so, connect the points to make the line that is the graph of an equation of the form $y = mx + b$ that has slope m.

   f. Note the location (ordered pair) that describes where the line intersects the y-axis.

Opening Exercise

Examine each of the graphs and their equations. Identify the coordinates of the point where the line intersects the y-axis. Describe the relationship between the point and the equation $y = mx + b$.

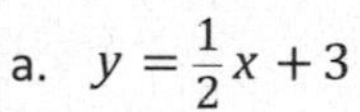

a. $y = \frac{1}{2}x + 3$

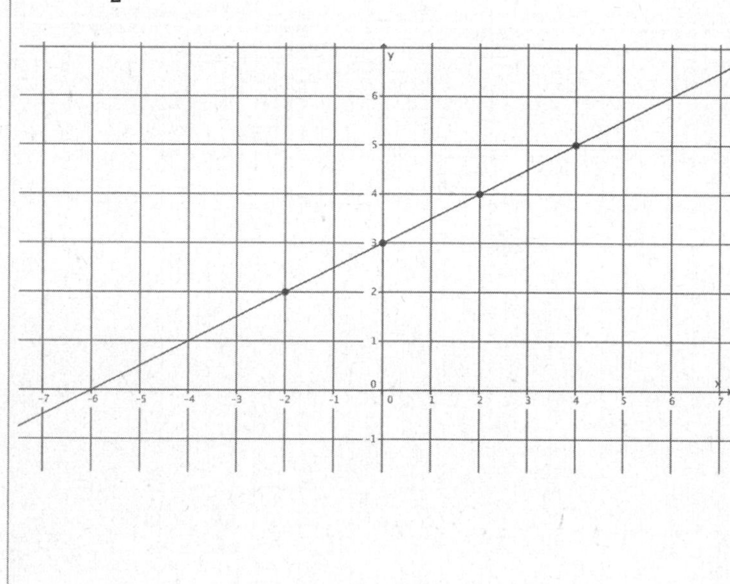

b. $y = -3x + 7$

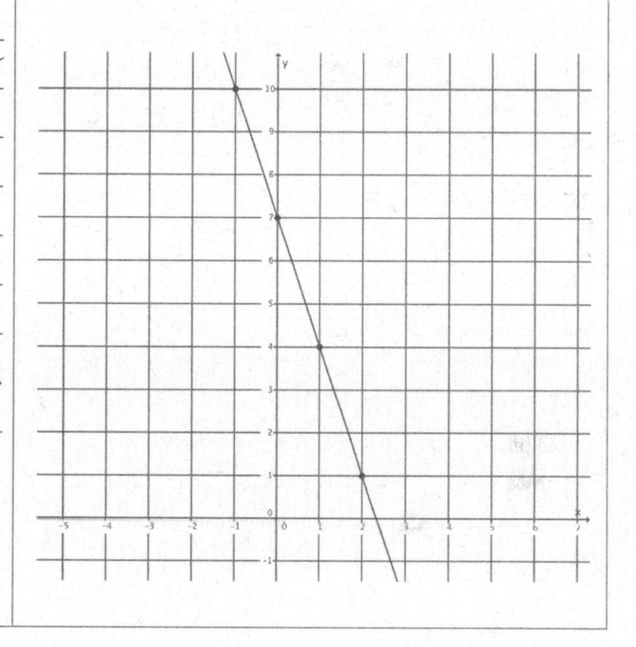

c. $y = -\frac{2}{3}x - 2$

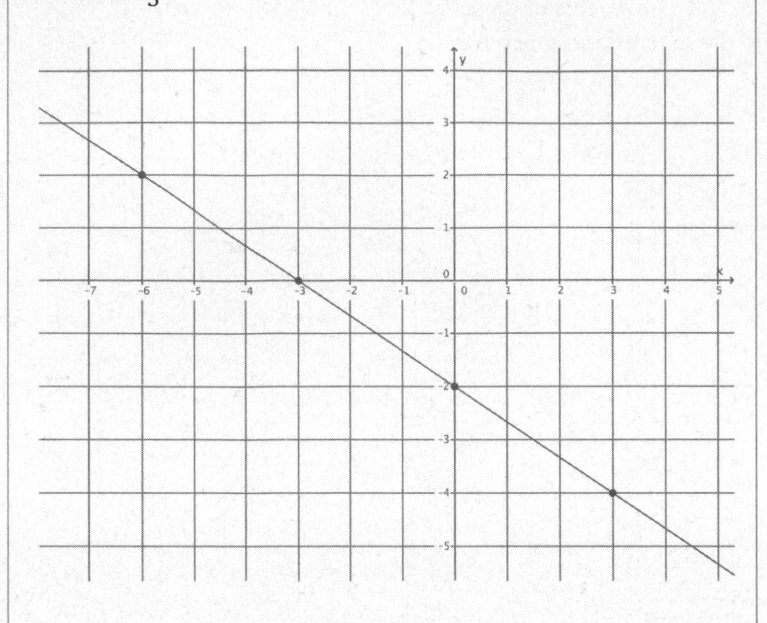

d. $y = 5x - 4$

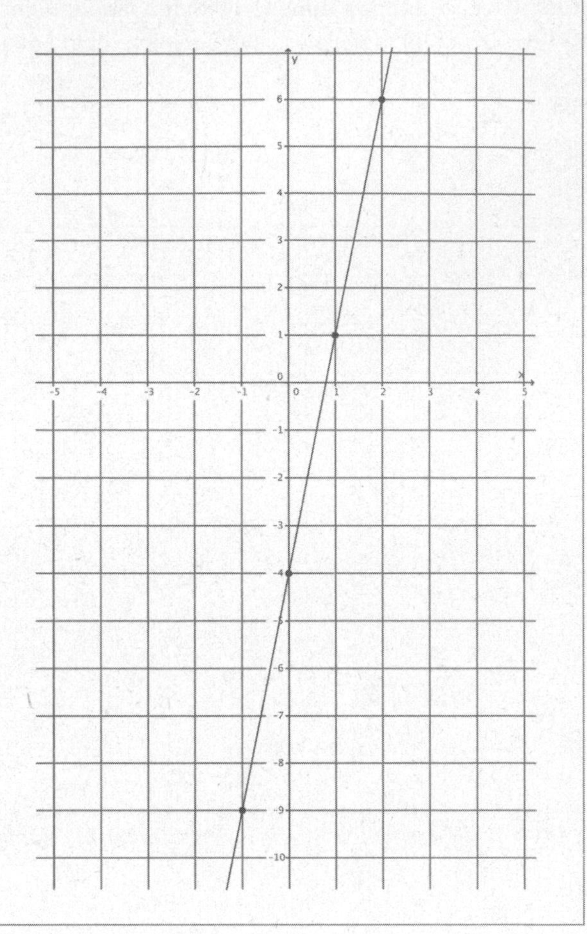

Example 1

Graph the equation $y = \frac{2}{3}x + 1$. Name the slope and y-intercept point.

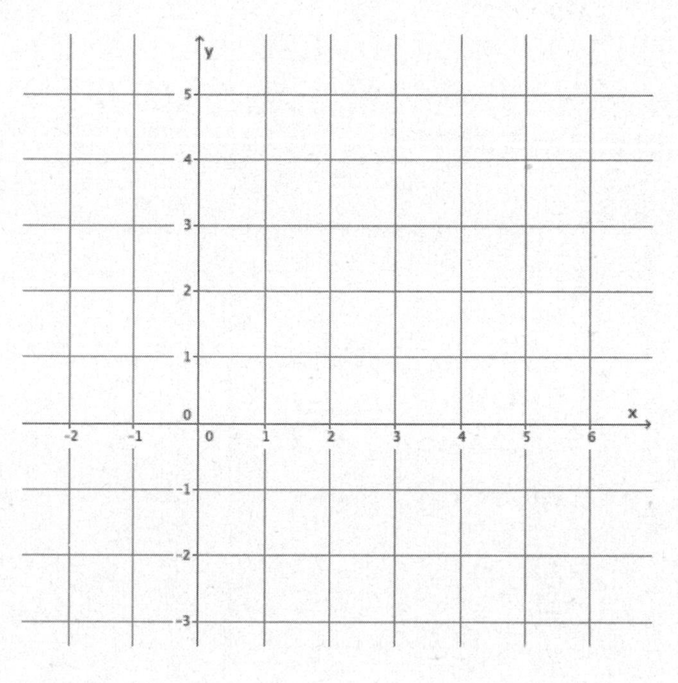

Example 2

Graph the equation $y = -\frac{3}{4}x - 2$. Name the slope and y-intercept point.

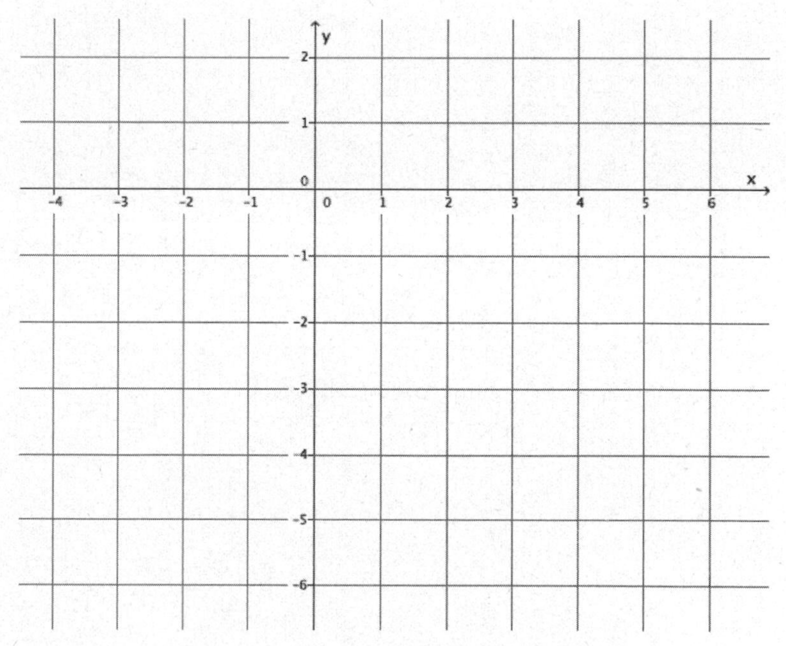

Example 3

Graph the equation $y = 4x - 7$. Name the slope and y-intercept point.

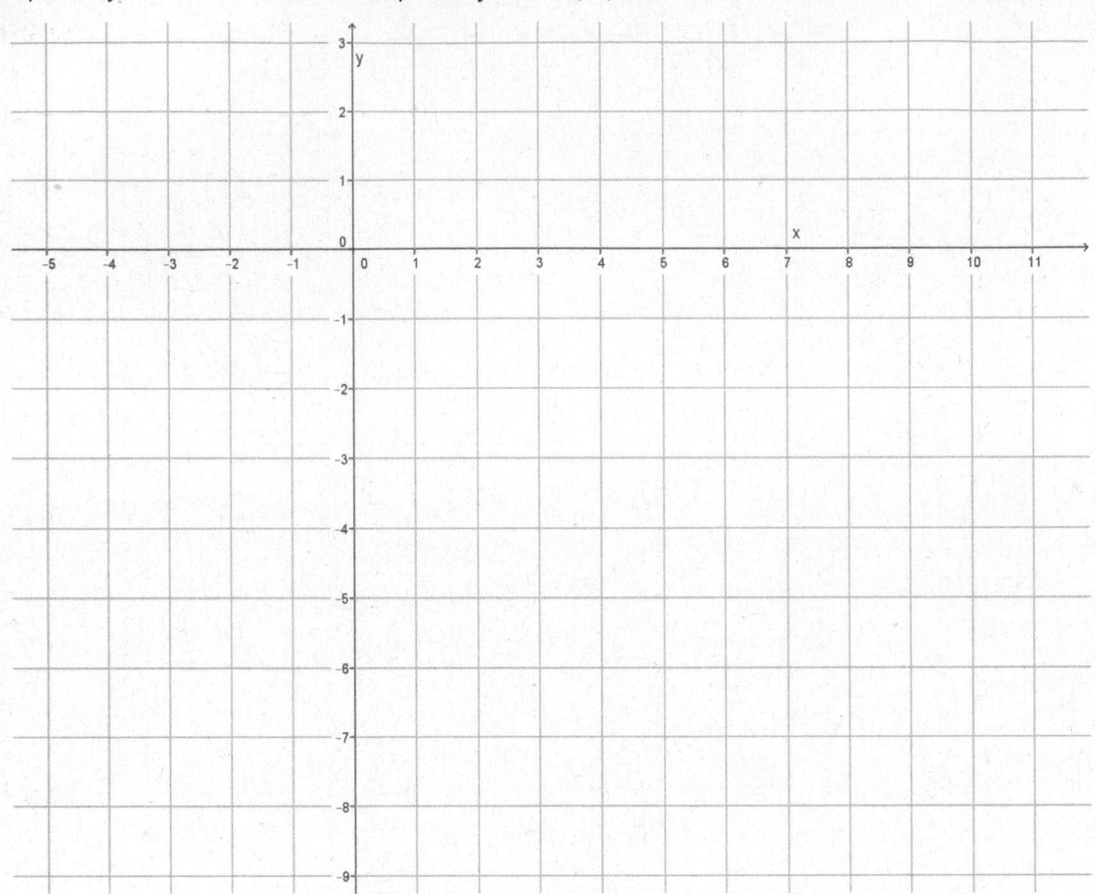

Exercises

1. Graph the equation $y = \frac{5}{2}x - 4$.

   a. Name the slope and the y-intercept point.

b. Graph the known point, and then use the slope to find a second point before drawing the line.

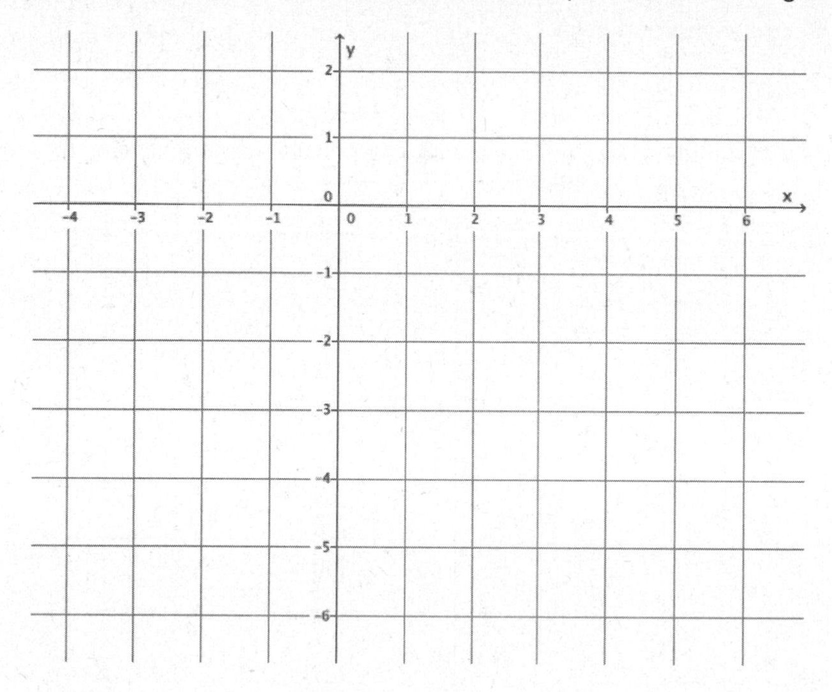

2. Graph the equation $y = -3x + 6$.

a. Name the slope and the y-intercept point.

b. Graph the known point, and then use the slope to find a second point before drawing the line.

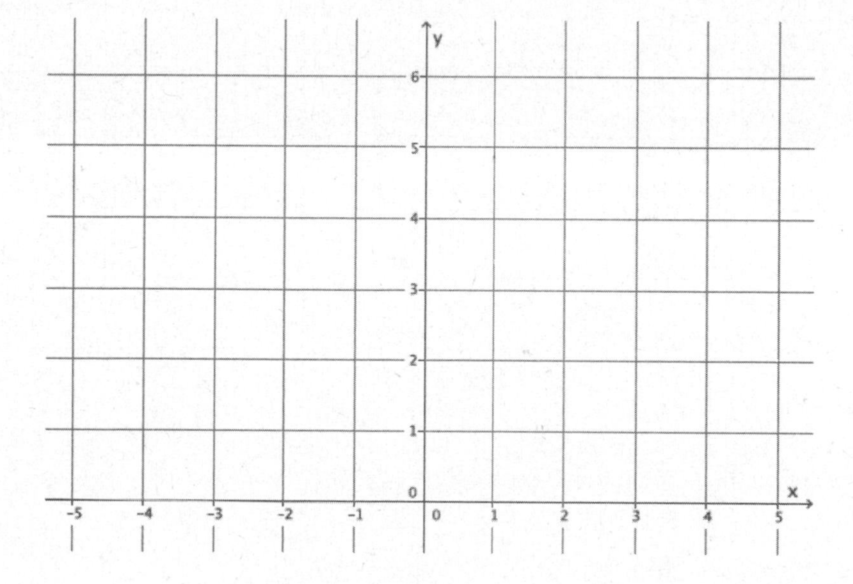

3. The equation $y = 1x + 0$ can be simplified to $y = x$. Graph the equation $y = x$.

   a. Name the slope and the y-intercept point.

   b. Graph the known point, and then use the slope to find a second point before drawing the line.

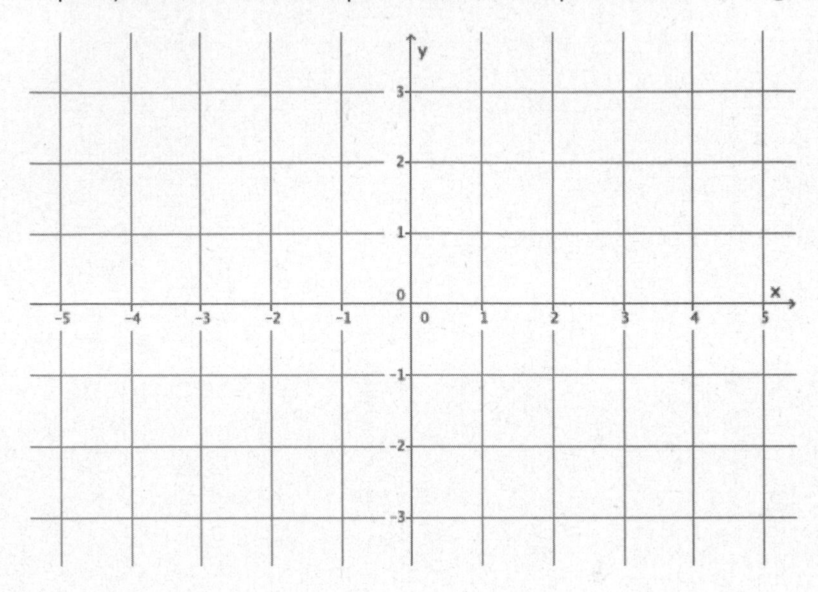

4. Graph the point $(0, 2)$.

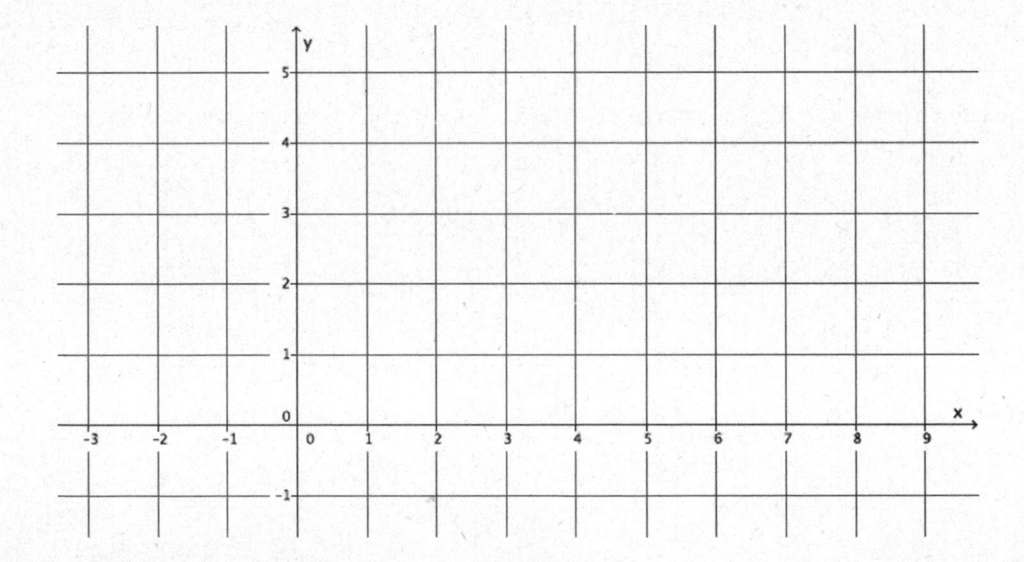

   a. Find another point on the graph using the slope, $m = \frac{2}{7}$.

   b. Connect the points to make the line.

c. Draw a different line that goes through the point $(0, 2)$ with slope $m = \frac{2}{7}$. What do you notice?

5. A bank put $\$10$ into a savings account when you opened the account. Eight weeks later, you have a total of $\$24$. Assume you saved the same amount every week.

   a. If y is the total amount of money in the savings account and x represents the number of weeks, write an equation in the form $y = mx + b$ that describes the situation.

   b. Identify the slope and the y-intercept point. What do these numbers represent?

   c. Graph the equation on a coordinate plane.

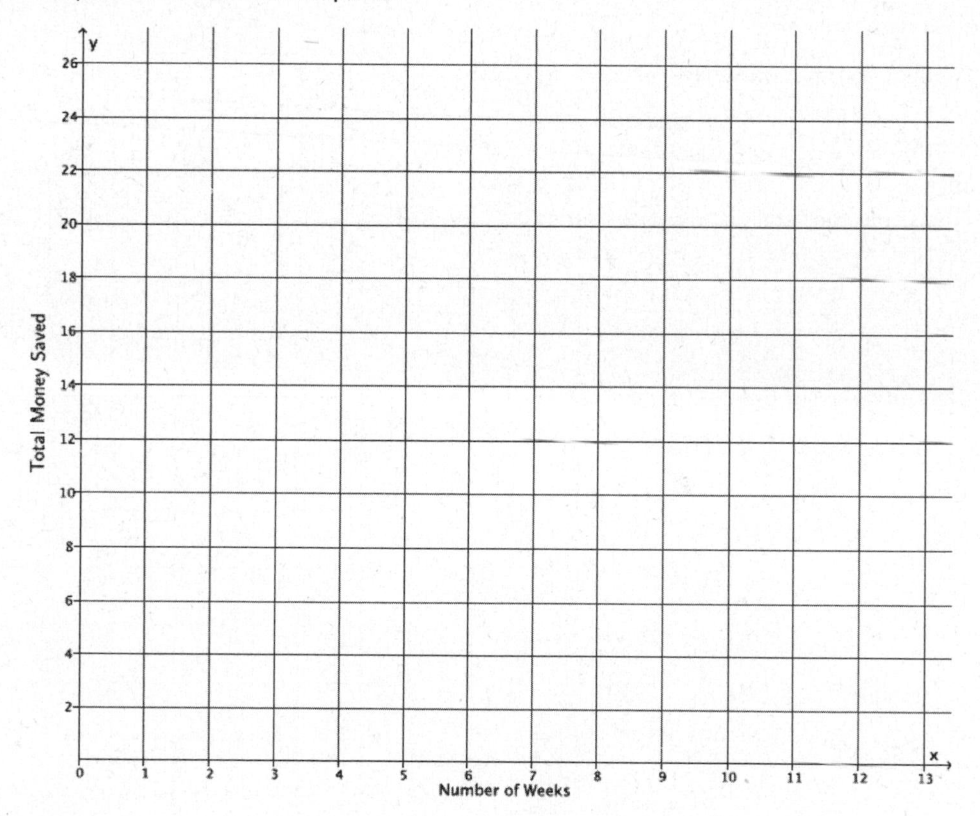

d. Could any other line represent this situation? For example, could a line through point $(0,10)$ with slope $\frac{7}{5}$ represent the amount of money you save each week? Explain.

6. A group of friends are on a road trip. After 120 miles, they stop to eat lunch. They continue their trip and drive at a constant rate of 50 miles per hour.

   a. Let y represent the total distance traveled, and let x represent the number of hours driven after lunch. Write an equation to represent the total number of miles driven that day.

   b. Identify the slope and the y-intercept point. What do these numbers represent?

   c. Graph the equation on a coordinate plane.

   d. Could any other line represent this situation? For example, could a line through point $(0, 120)$ with slope 75 represent the total distance the friends drive? Explain.

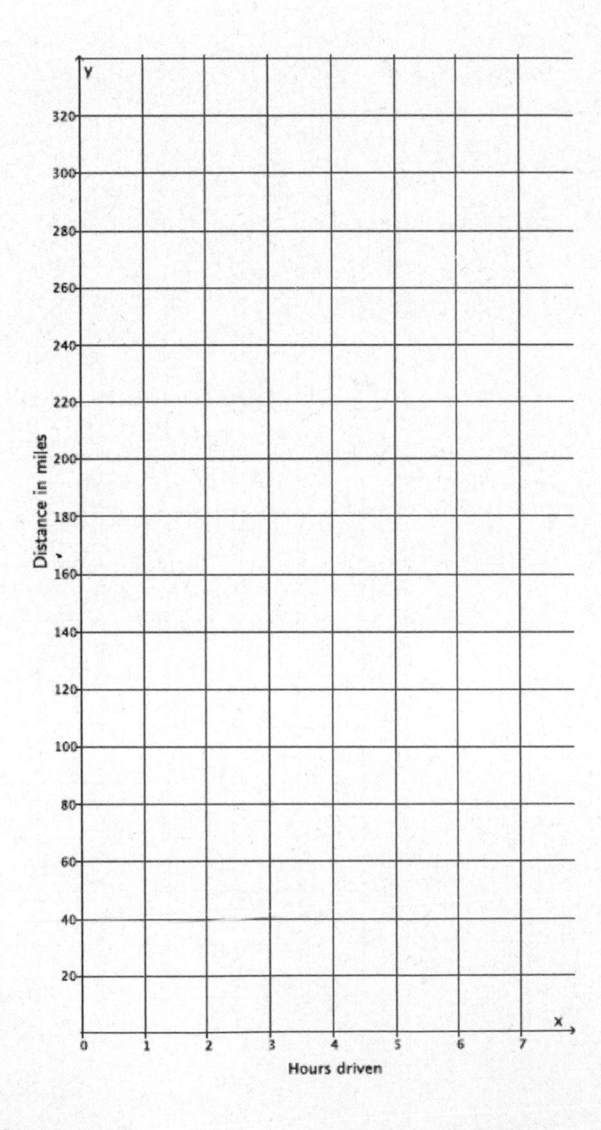

Lesson Summary

The equation $y = mx + b$ is in slope-intercept form. The number m represents the slope of the graph, and the point $(0, b)$ is the location where the graph of the line intersects the y-axis.

To graph a line from the slope-intercept form of a linear equation, begin with the known point, $(0, b)$, and then use the slope to find a second point. Connect the points to graph the equation.

There is only one line passing through a given point with a given slope.

2. Student graph of $y = -\frac{3}{5}x - 1$:

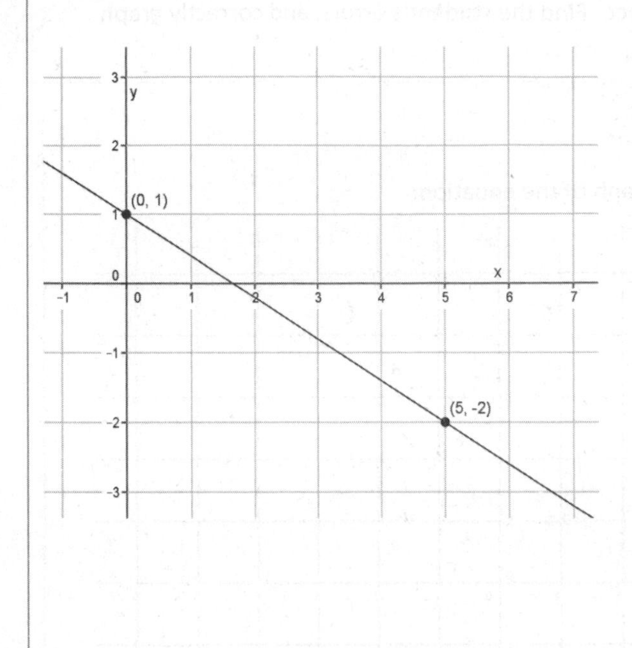

Error:

Correct graph of the equation:

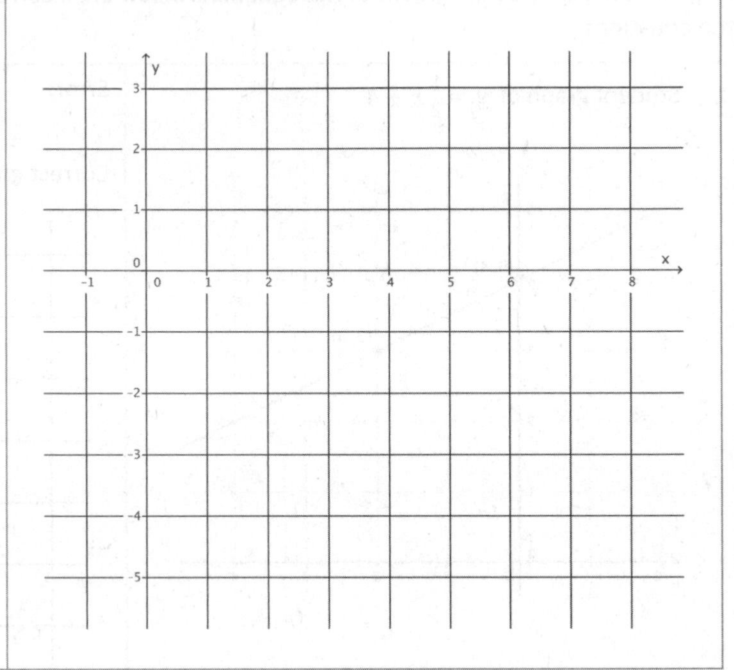

Graph each equation on a separate pair of x- and y-axes. Students need graph paper to complete the Problem Set.

1. Graph the equation $y = \frac{4}{3}x + 2$.

I know the equation is in slope-intercept form, $y = mx + b$, the number m represents the slope of the graph, and the point $(0, b)$ is the location where the graph of the line intersects the y-axis.

   a. Name the slope and the y-intercept point.

   The slope is $m = \frac{4}{3}$, and the y-intercept point is $(0, 2)$.

   b. Graph the known point, and then use the slope to find a second point before drawing the line.

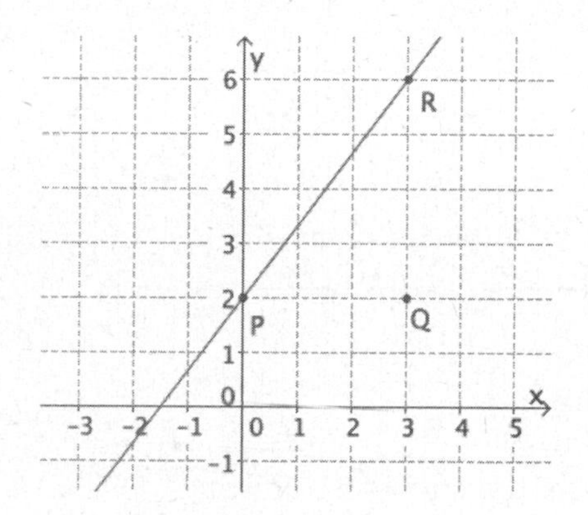

I know that $m = \frac{|QR|}{|PQ|}$. Since $|PQ| = 3$, I need to go 3 units from the right of point P, the y-intercept, to find point Q. Since $|QR| = 4$ and the slope is positive, I need to go up 4 units from point Q to find point R.

2. Graph the equation $y = -\frac{1}{3}x + 5$.

   a. Name the slope and the y-intercept point.

   The slope is $m = -\frac{1}{3}$, and the y-intercept point is $(0, 5)$.

b. Graph the known point, and then use the slope to find a second point before drawing the line.

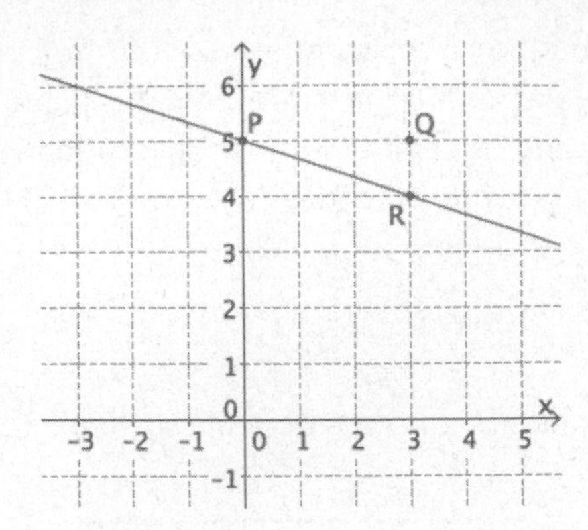

I need to go three units to the right of point P and mark point Q. Since the slope is negative, I need to go down 1 unit from point Q to find point R.

3. Graph the equation $y = \frac{1}{4}x$.

Rewriting the equation in slope-intercept form, $y = \frac{1}{4}x + 0$, helps me to see the y-intercept point is $(0, 0)$.

a. Name the slope and the y-intercept point.

The slope is $m = \frac{1}{4}$***, and the y-intercept point is*** $(0, 0)$.

b. Graph the known point, and then use the slope to find a second point before drawing the line.

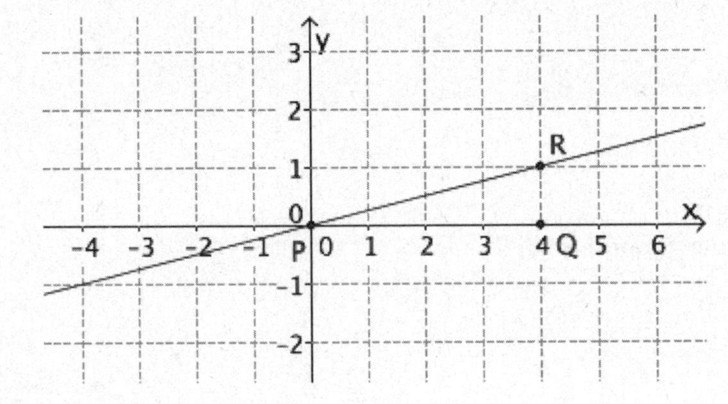

4. Graph the equation $2x + 2y = 2$.

a. Name the slope and the y-intercept point.

I need to rewrite the equation in slope-intercept form to help me name the slope and y-intercept point more easily.

$$2x + 2y = 2$$
$$2x - 2x + 2y = -2x + 2$$
$$2y = -2x + 2$$
$$\frac{2}{2}y = -\frac{2}{2}x + \frac{2}{2}$$
$$y = -x + 1$$

The slope -1 is equivalent to the fraction $-\frac{1}{1}$.

The slope is m $= -1$***, and the y-intercept point is*** $(0, 1)$.

b. Graph the known point, and then use the slope to find a second point before drawing the line.

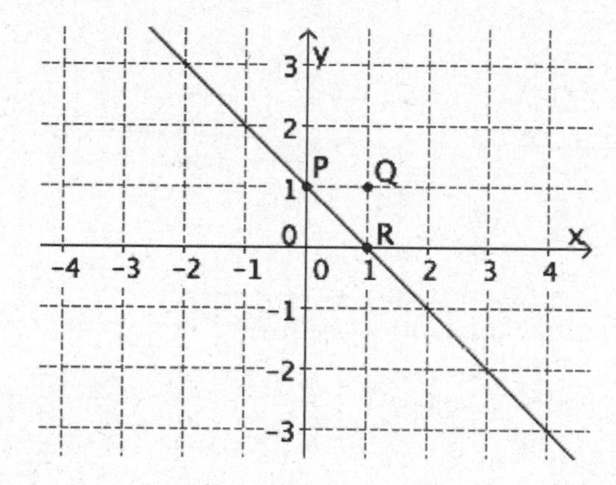

Graph each equation on a separate pair of x- and y-axes.

1. Graph the equation $y = \frac{4}{5}x - 5$.
   a. Name the slope and the y-intercept point.
   b. Graph the known point, and then use the slope to find a second point before drawing the line.

2. Graph the equation $y = x + 3$.
   a. Name the slope and the y-intercept point.
   b. Graph the known point, and then use the slope to find a second point before drawing the line.

3. Graph the equation $y = -\frac{4}{3}x + 4$.
   a. Name the slope and the y-intercept point.
   b. Graph the known point, and then use the slope to find a second point before drawing the line.

4. Graph the equation $y = \frac{5}{2}x$.
   a. Name the slope and the y-intercept point.
   b. Graph the known point, and then use the slope to find a second point before drawing the line.

5. Graph the equation $y = 2x - 6$.
   a. Name the slope and the y-intercept point.
   b. Graph the known point, and then use the slope to find a second point before drawing the line.

6. Graph the equation $y = -5x + 9$.
   a. Name the slope and the y-intercept point.
   b. Graph the known point, and then use the slope to find a second point before drawing the line.

7. Graph the equation $y = \frac{1}{3}x + 1$.
   a. Name the slope and the y-intercept point.
   b. Graph the known point, and then use the slope to find a second point before drawing the line.

8. Graph the equation $5x + 4y = 8$. (Hint: Transform the equation so that it is of the form $y = mx + b$.)
   a. Name the slope and the y-intercept point.
   b. Graph the known point, and then use the slope to find a second point before drawing the line.

9. Graph the equation $-2x + 5y = 30$.
   a. Name the slope and the y-intercept point.
   b. Graph the known point, and then use the slope to find a second point before drawing the line.

10. Let l and l' be two lines with the same slope m passing through the same point P. Show that there is only one line with a slope m, where $m < 0$, passing through the given point P. Draw a diagram if needed.

Exercises

THEOREM: The graph of a linear equation $y = mx + b$ is a non-vertical line with slope m and passing through $(0, b)$, where b is a constant.

1. Prove the theorem by completing parts (a)–(c). Given two distinct points, P and Q, on the graph of $y = mx + b$, and let l be the line passing through P and Q. You must show the following:

   (1) Any point on the graph of $y = mx + b$ is on line l, and

   (2) Any point on the line l is on the graph of $y = mx + b$.

   a. Proof of (1): Let R be any point on the graph of $y = mx + b$. Show that R is on l. Begin by assuming it is not. Assume the graph looks like the diagram below where R is on l'.

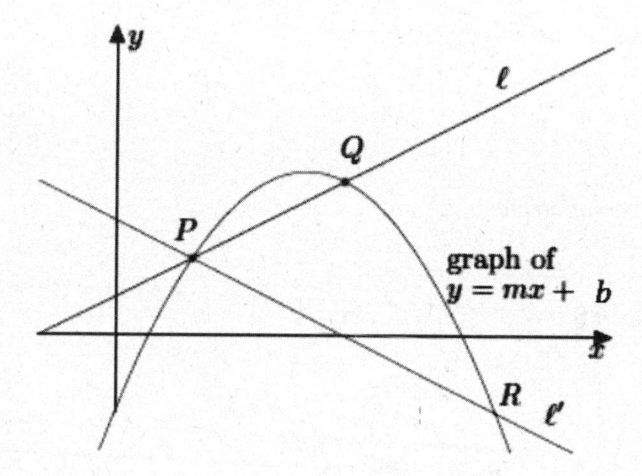

   What is the slope of line l?

What is the slope of line l'?

What can you conclude about lines l and l'? Explain.

b. Proof of (2): Let S be any point on line l, as shown.

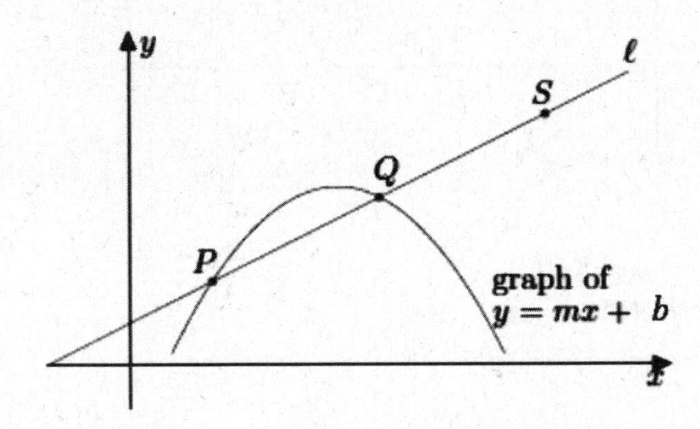

Show that S is a solution to $y = mx + b$. Hint: Use the point $(0, b)$.

c. Now that you have shown that any point on the graph of $y = mx + b$ is on line l in part (a), and any point on line l is on the graph of $y = mx + b$ in part (b), what can you conclude about the graphs of linear equations?

2. Use $x = 4$ and $x = -4$ to find two solutions to the equation $x + 2y = 6$. Plot the solutions as points on the coordinate plane, and connect the points to make a line.

   a. Identify two other points on the line with integer coordinates. Verify that they are solutions to the equation $x + 2y = 6$.

   b. When $x = 1$, what is the value of y? Does this solution appear to be a point on the line?

   c. When $x = -3$, what is the value of y? Does this solution appear to be a point on the line?

   d. Is the point $(3, 2)$ on the line?

   e. Is the point $(3, 2)$ a solution to the linear equation $x + 2y = 6$?

3. Use $x = 4$ and $x = 1$ to find two solutions to the equation $3x - y = 9$. Plot the solutions as points on the coordinate plane, and connect the points to make a line.

   a. Identify two other points on the line with integer coordinates. Verify that they are solutions to the equation $3x - y = 9$.

   b. When $x = 4.5$, what is the value of y? Does this solution appear to be a point on the line?

   c. When $x = \frac{1}{2}$, what is the value of y? Does this solution appear to be a point on the line?

   d. Is the point $(2, 4)$ on the line?

   e. Is the point $(2, 4)$ a solution to the linear equation $3x - y = 9$?

4. Use $x = 3$ and $x = -3$ to find two solutions to the equation $2x + 3y = 12$. Plot the solutions as points on the coordinate plane, and connect the points to make a line.

   a. Identify two other points on the line with integer coordinates. Verify that they are solutions to the equation $2x + 3y = 12$.

b. When $x = 2$, what is the value of y? Does this solution appear to be a point on the line?

c. When $x = -2$, what is the value of y? Does this solution appear to be a point on the line?

d. Is the point $(8, -3)$ on the line?

e. Is the point $(8, -3)$ a solution to the linear equation $2x + 3y = 12$?

5. Use $x = 4$ and $x = -4$ to find two solutions to the equation $x - 2y = 8$. Plot the solutions as points on the coordinate plane, and connect the points to make a line.

   a. Identify two other points on the line with integer coordinates. Verify that they are solutions to the equation $x - 2y = 8$.

   b. When $x = 7$, what is the value of y? Does this solution appear to be a point on the line?

c. When $x = -3$, what is the value of y? Does this solution appear to be a point on the line?

d. Is the point $(-2, -3)$ on the line?

e. Is the point $(-2, -3)$ a solution to the linear equation $x - 2y = 8$?

6. Based on your work in Exercises 2–5, what conclusions can you draw about the points on a line and solutions to a linear equation?

7. Based on your work in Exercises 2–5, will a point that is not a solution to a linear equation be a point on the graph of a linear equation? Explain.

8. Based on your work in Exercises 2–5, what conclusions can you draw about the graph of a linear equation?

9. Graph the equation $-3x + 8y = 24$ using intercepts.

10. Graph the equation $x - 6y = 15$ using intercepts.

11. Graph the equation $4x + 3y = 21$ using intercepts.

Lesson Summary

The graph of a linear equation is a line. A linear equation can be graphed using two-points: the x-intercept point and the y-intercept point.

Example:

Graph the equation: $2x + 3y = 9$.

Replace x with zero, and solve for y to determine the y-intercept point.

$$2(0) + 3y = 9$$
$$3y = 9$$
$$y = 3$$

The y-intercept point is at $(0, 3)$.

Replace y with zero, and solve for x to determine the x-intercept point.

$$2x + 3(0) = 9$$
$$2x = 9$$
$$x = \frac{9}{2}$$

The x-intercept point is at $(\frac{9}{2}, 0)$.

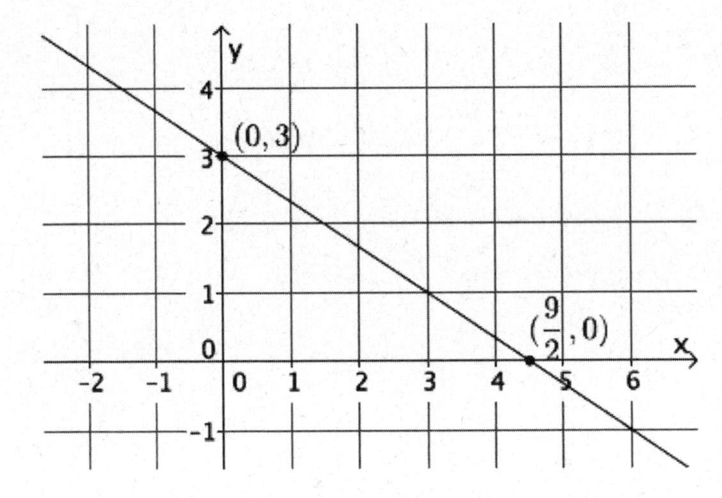

Name ____________________ Date ____________

1. Graph the equation $y = \frac{5}{4}x - 10$ using the y-intercept point and slope.

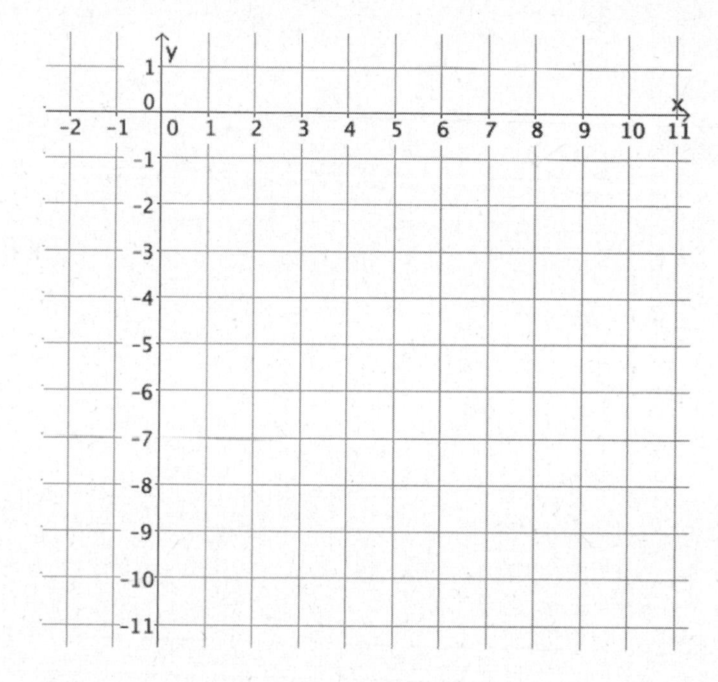

2. Graph the equation $5x - 4y = 40$ using intercepts.

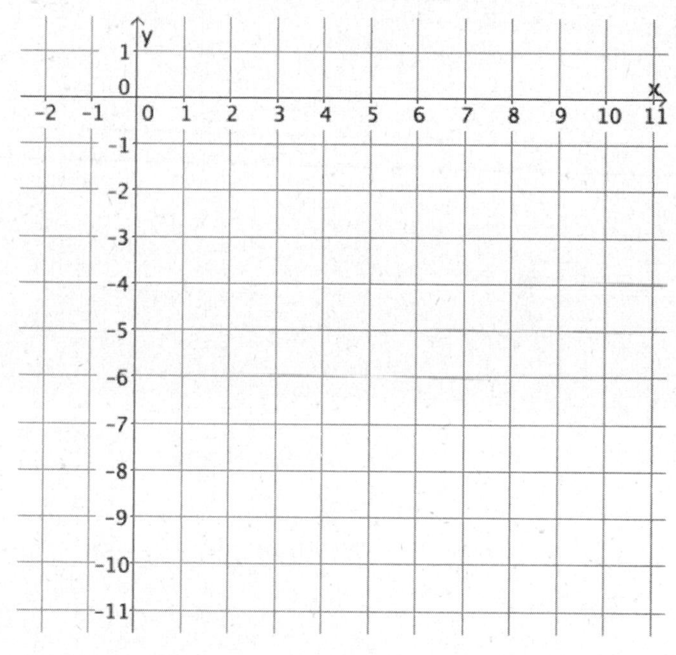

3. What can you conclude about the equations $y = \frac{5}{4}x - 10$ and $5x - 4y = 40$?

Students need graph paper to complete the Problem Set.

1. Graph the equation: $y = \frac{1}{2}x - 2$.

This is a linear equation in slope-intercept form. I will use the slope $m = \frac{1}{2}$ and the y-intercept point $(0, -2)$ to graph the linear equation.

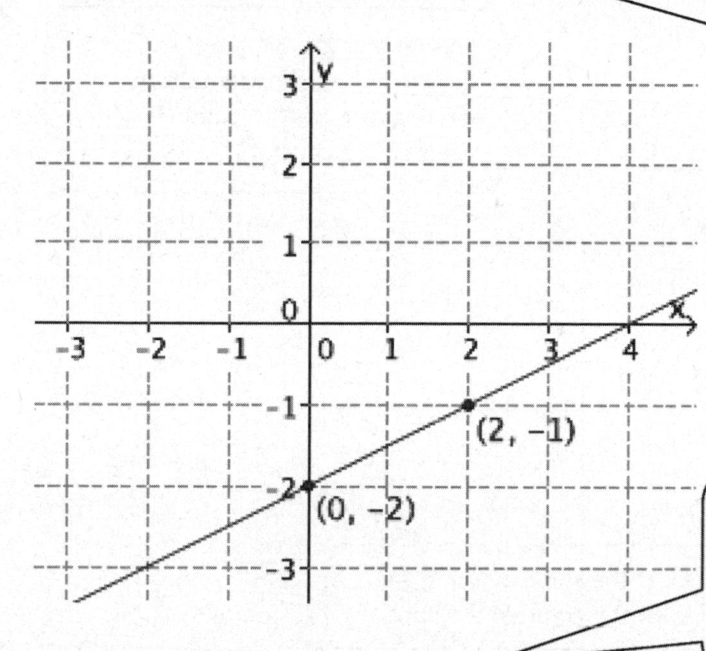

2. Graph the equation: $4x + 8y = 16$.

This is a linear equation in standard form. I will find the y-intercept point by replacing the x with 0. I will find the x-intercept point by replacing the y with 0.

$$4(0) + 8y = 16$$
$$8y = 16$$
$$y = 2$$

The y-intercept point is $(0, 2)$.

$$4x + 8(0) = 16$$
$$4x = 16$$
$$x = 4$$

The x-intercept point is $(4, 0)$.

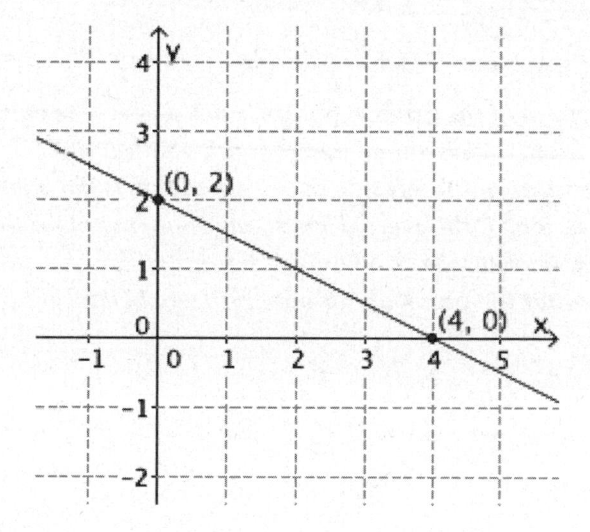

3. Graph the equation: $y = -2$. What is the slope of the graph of this line?

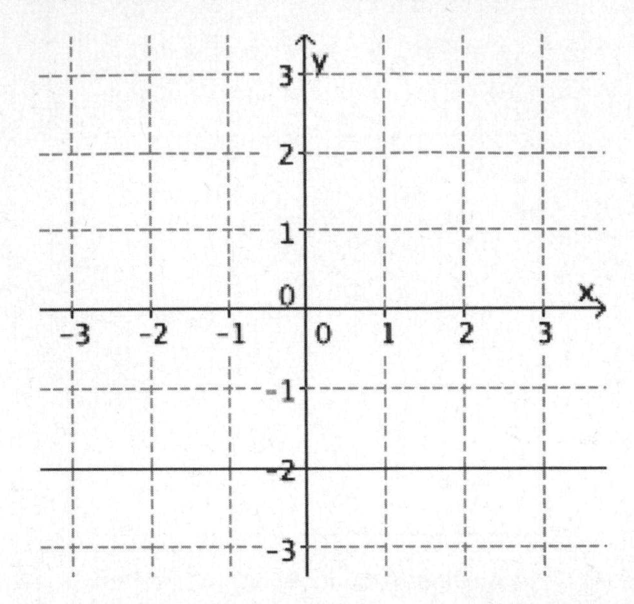

I remember that equations of the form $y = b$ are horizontal lines passing through the point $(0, b)$ where b is a constant.

The slope of this line is zero.

I can calculate the slope by using any two points on the graph of the line.

4. Is the graph of $x^2 - 6y = 11$ a line? Explain.

The graph of the given equation is not a line. The equation $6x^2 - 6y = 11$ ***is not a linear equation because the expression on the left side of the equal sign is not a linear expression. If this were a linear equation, then I would be sure that it graphs as a line, but because it is not, I am not sure what the graph of this equation would look like.***

Linear expressions are constants like -1 or 5. Linear expressions can be a product of constants and an x like $5x$ or $-2x$, or a product of constants and a y like $9y$ or $-11y$.

Graph each of the equations in the Problem Set on a different pair of x- and y-axes.

1. Graph the equation: $y = -6x + 12$.

2. Graph the equation: $9x + 3y = 18$.

3. Graph the equation: $y = 4x + 2$.

4. Graph the equation: $y = -\frac{5}{7}x + 4$.

5. Graph the equation: $\frac{3}{4}x + y = 8$.

6. Graph the equation: $2x - 4y = 12$.

7. Graph the equation: $y = 3$. What is the slope of the graph of this line?

8. Graph the equation: $x = -4$. What is the slope of the graph of this line?

9. Is the graph of $4x + 5y = \frac{3}{7}$ a line? Explain.

10. Is the graph of $6x^2 - 2y = 7$ a line? Explain.

Opening Exercise

Figure 1

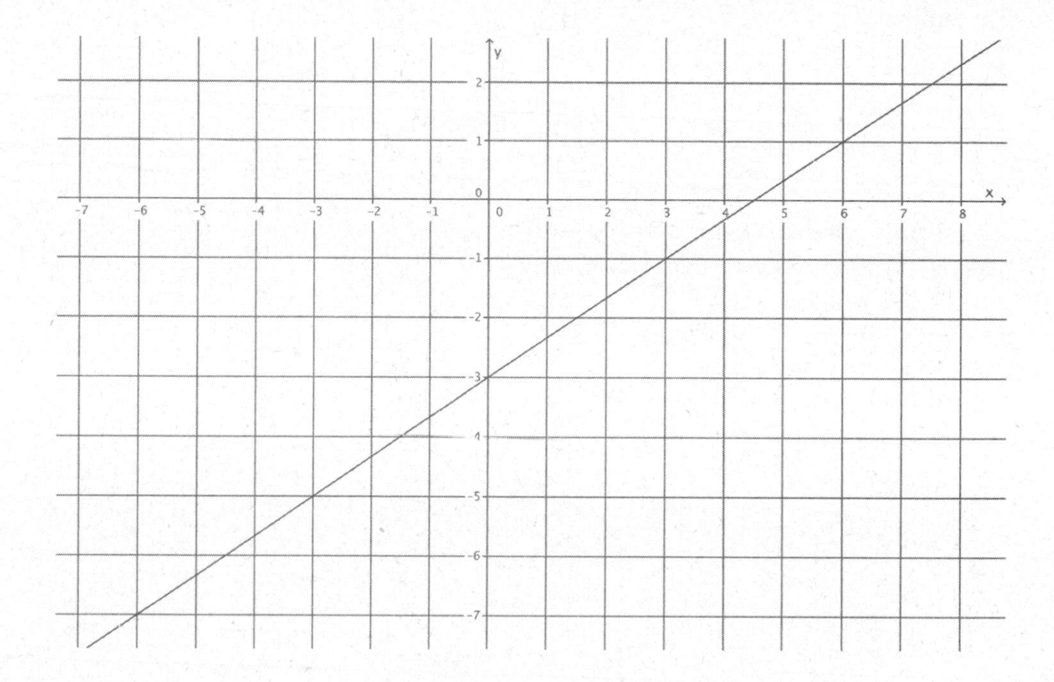

Figure 2

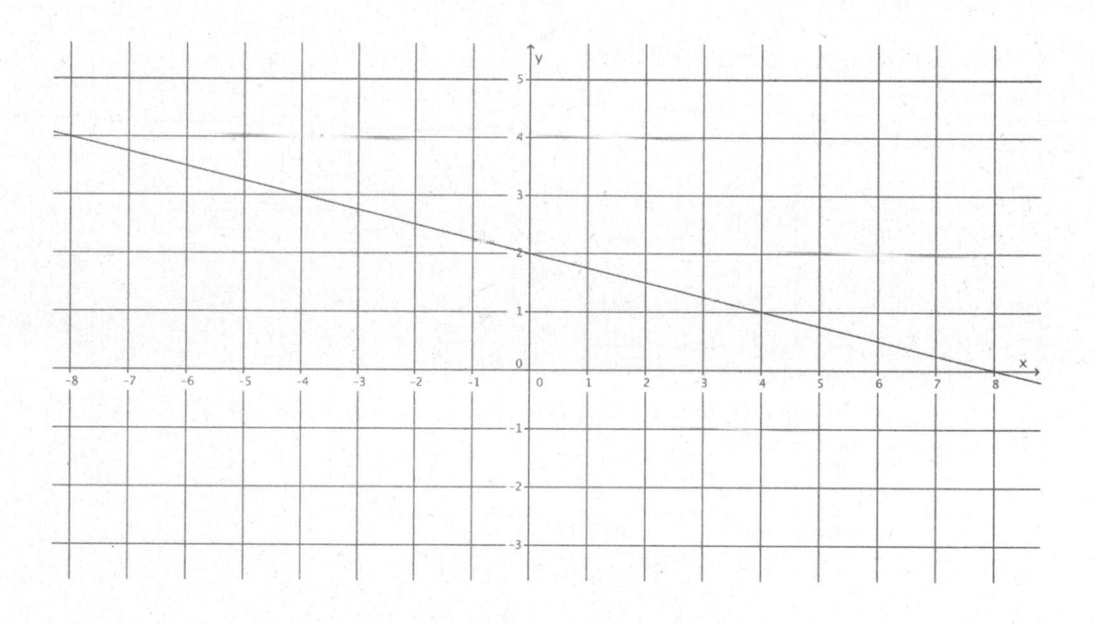

Exercises

1. Write the equation that represents the line shown.

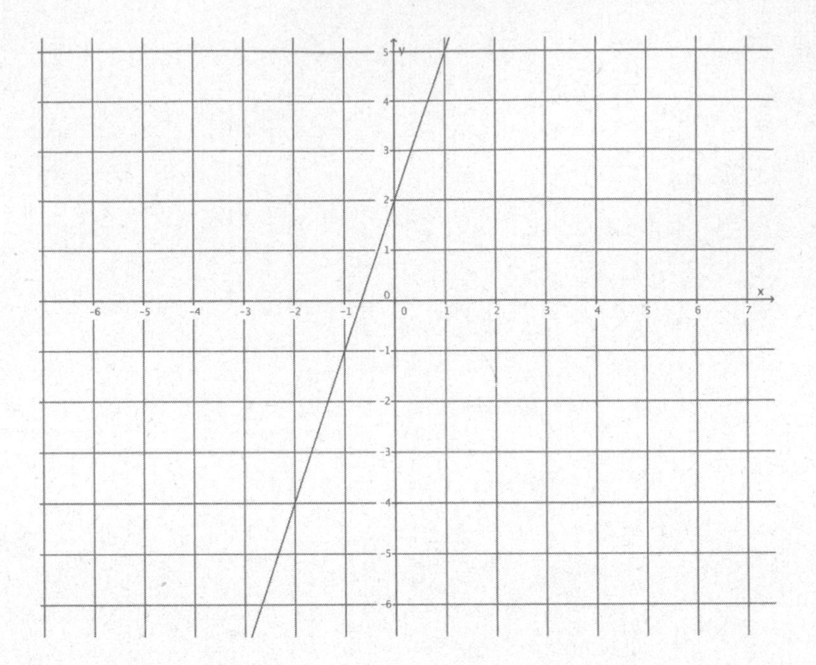

   Use the properties of equality to change the equation from slope-intercept form, $y = mx + b$, to standard form, $ax + by = c$, where a, b, and c are integers, and a is not negative.

2. Write the equation that represents the line shown.

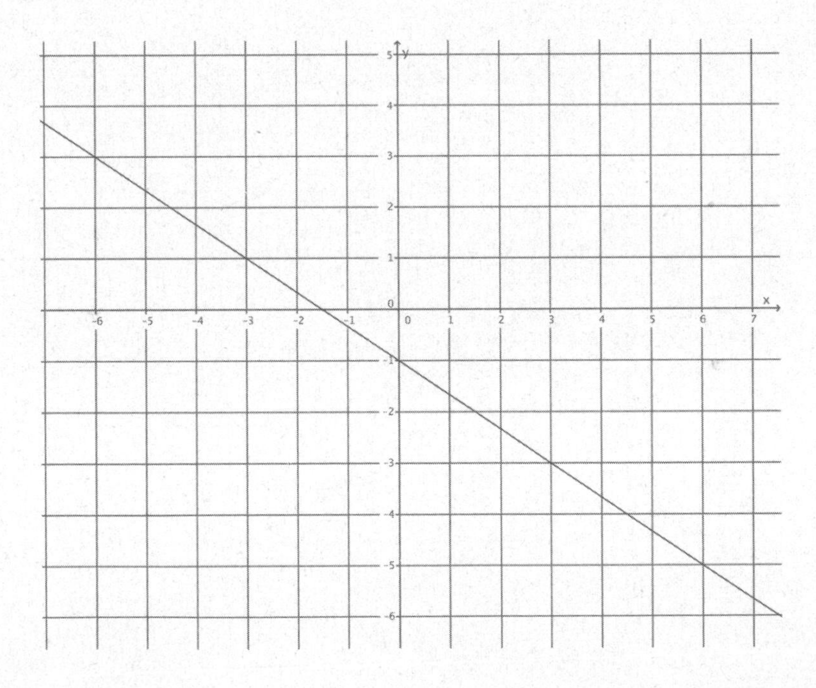

   Use the properties of equality to change the equation from slope-intercept form, $y = mx + b$, to standard form, $ax + by = c$, where a, b, and c are integers, and a is not negative.

3. Write the equation that represents the line shown.

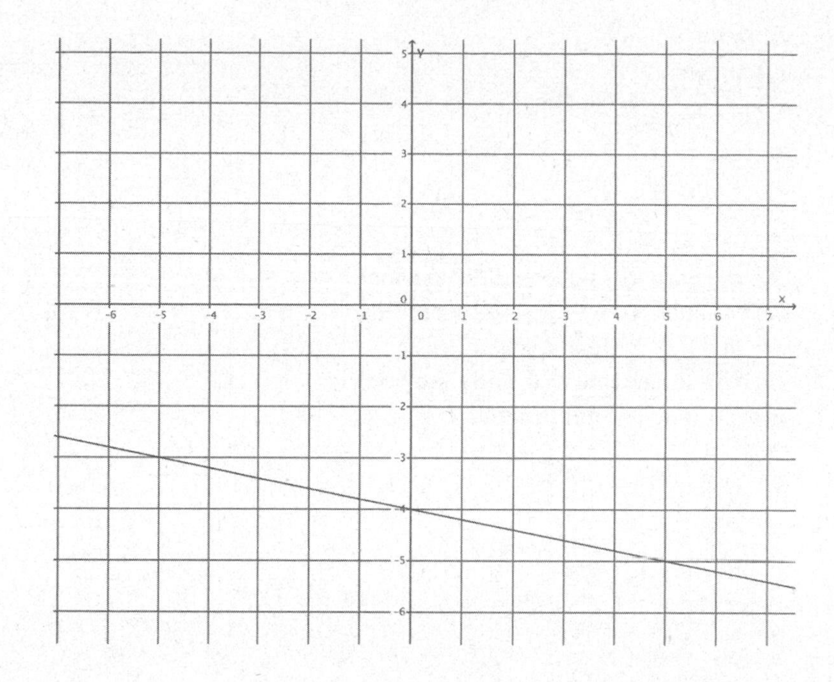

Use the properties of equality to change the equation from slope-intercept form, $y = mx + b$, to standard form, $ax + by = c$, where a, b, and c are integers, and a is not negative.

4. Write the equation that represents the line shown.

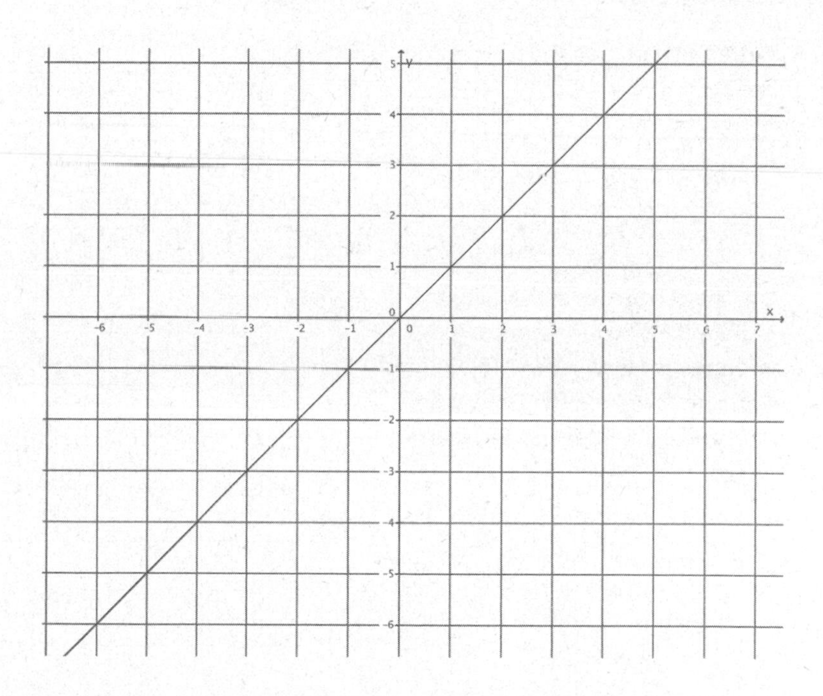

Use the properties of equality to change the equation from slope-intercept form, $y = mx + b$, to standard form, $ax + by = c$, where a, b, and c are integers, and a is not negative.

5. Write the equation that represents the line shown.

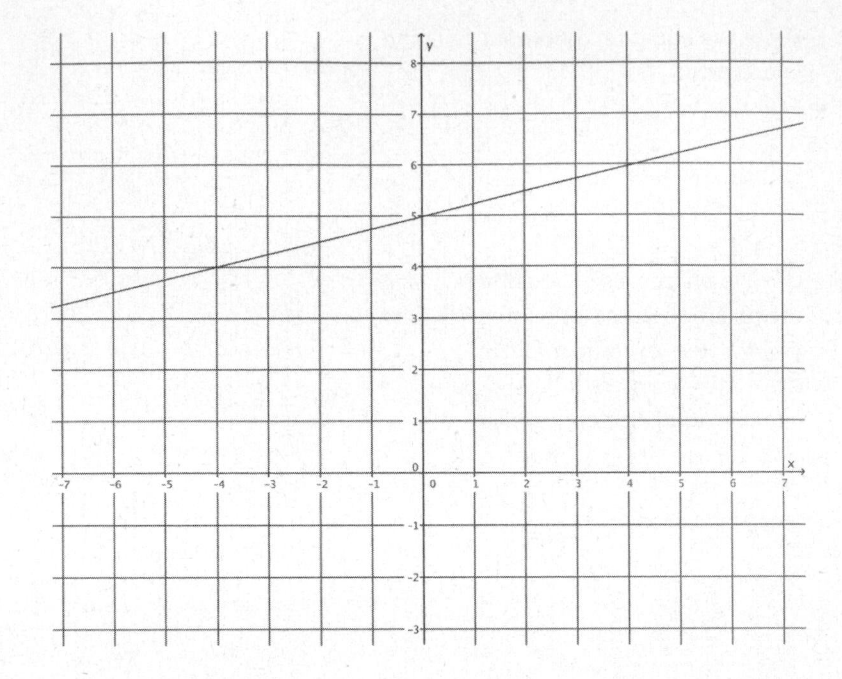

Use the properties of equality to change the equation from slope-intercept form, $y = mx + b$, to standard form, $ax + by = c$, where a, b, and c are integers, and a is not negative.

6. Write the equation that represents the line shown.

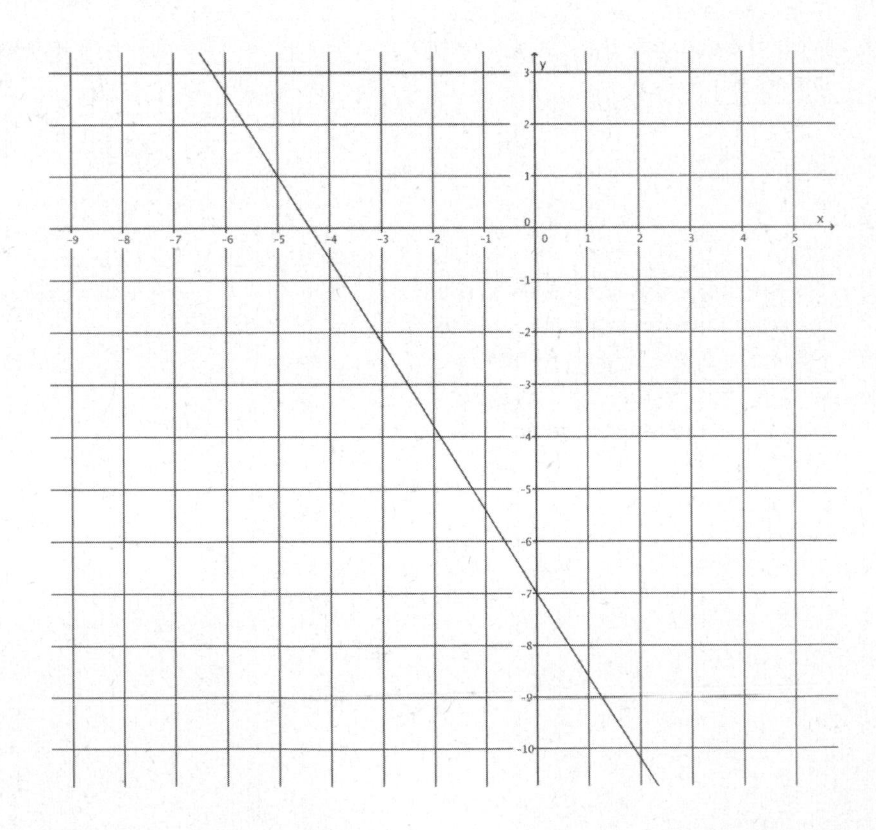

Use the properties of equality to change the equation from slope-intercept form, $y = mx + b$, to standard form, $ax + by = c$, where a, b, and c are integers, and a is not negative.

Lesson Summary

Write the equation of a line by determining the y-intercept point, $(0, b)$, and the slope, m, and replacing the numbers b and m into the equation $y = mx + b$.

Example:

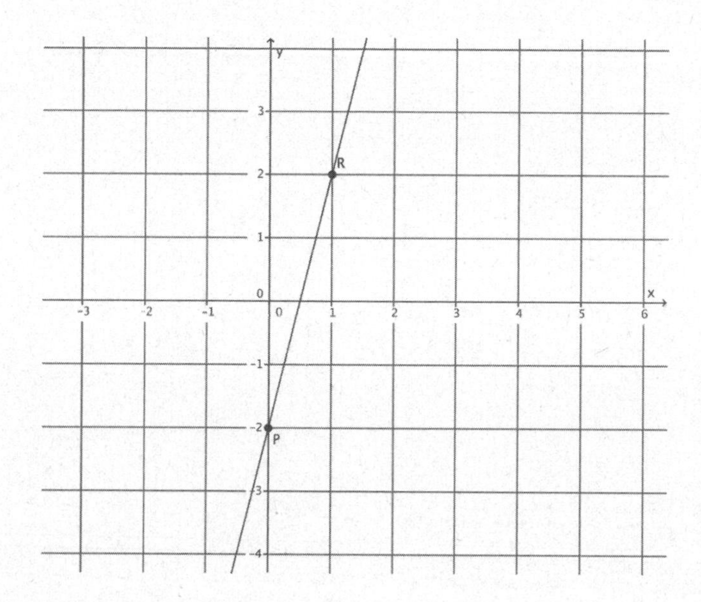

The y-intercept point of this graph is $(0, -2)$.

The slope of this graph is $m = \frac{4}{1} = 4$.

The equation that represents the graph of this line is $y = 4x - 2$.

Use the properties of equality to change the equation from slope-intercept form, $y = mx + b$, to standard form, $ax + by = c$, where a, b, and c are integers, and a is not negative.

Name ______________________________ Date ______________

1. Write an equation in slope-intercept form that represents the line shown.

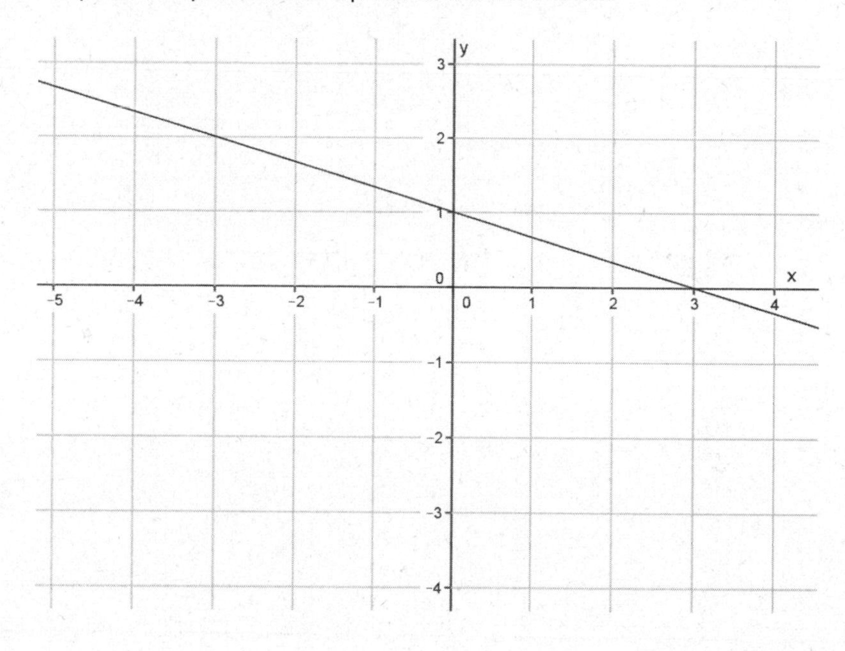

2. Use the properties of equality to change the equation you wrote for Problem 1 from slope-intercept form, $y = mx + b$, to standard form, $ax + by = c$, where a, b, and c are integers, and a is not negative.

3. Write an equation in slope-intercept form that represents the line shown.

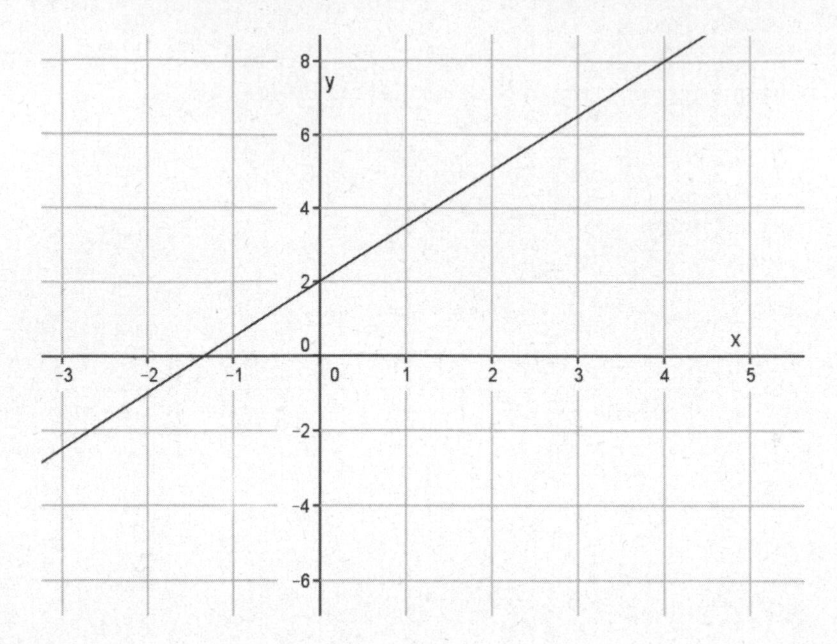

4. Use the properties of equality to change the equation you wrote for Problem 3 from slope-intercept form, $y = mx + b$, to standard form, $ax + by = c$, where a, b, and c are integers, and a is not negative.

1. Write the equation that represents the line shown.

$$y = -\frac{1}{2}x + 2$$

I identified point P as the y-intercept, which is $(0, 2)$. I can use any point on the graph for point R, so I will use $(-4, 4)$. Point Q will be $(-4, 2)$. This will help me find the slope of $-\frac{2}{4}$, which is equivalent to $-\frac{1}{2}$. I will substitute the information into the slope-intercept form of the equation.

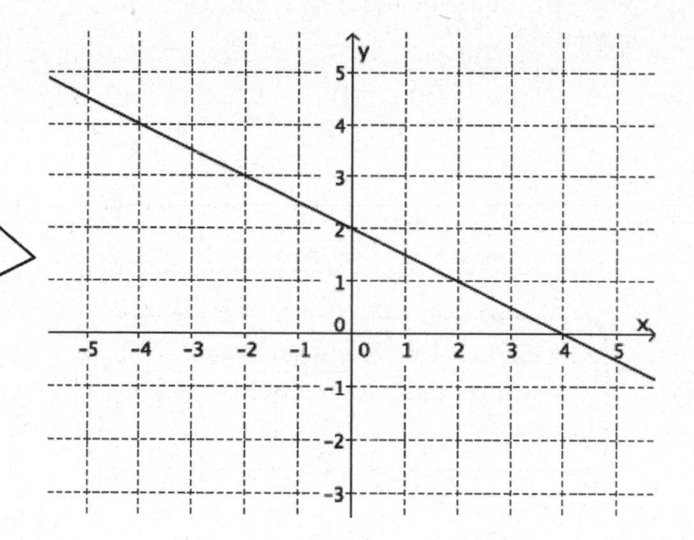

a. Use the properties of equality to change the equation from slope-intercept form, $y = mx + b$, to standard form, $ax + by = c$, where a, b, and c are integers, and a is not negative.

$$y = -\frac{1}{2}x + 2$$

$$\left(y = -\frac{1}{2}x + 2\right)2$$

$$2y = -x + 4$$

$$x + 2y = -x + x + 4$$

$$x + 2y = 4$$

What number can I multiply the equation by so that $-\frac{1}{2}$ will become an integer?

2. Write the equation that represents the line shown.

I need to calculate the slope and determine the y-intercept like I did in Problem 1.

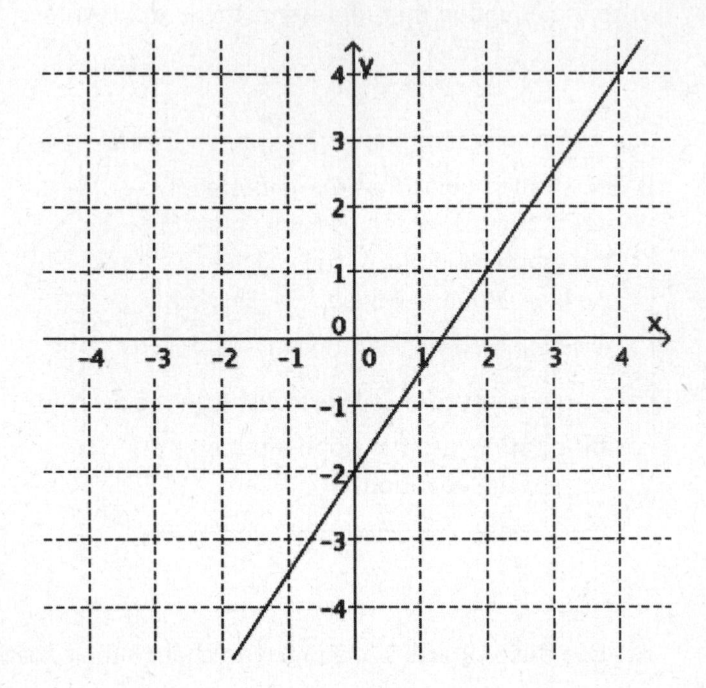

a. Use the properties of equality to change the equation from slope-intercept form, $y = mx + b$, to standard form, $ax + by = c$, where a, b, and c are integers, and a is not negative.

$$y = \frac{3}{2}x - 2$$
$$\left(y = \frac{3}{2}x - 2\right)2$$
$$2y = 3x - 4$$
$$-3x + 2y = 3x - 3x - 4$$
$$-3x + 2y = -4$$
$$-1(-3x + 2y = -4)$$
$$3x - 2y = 4$$

I need to multiply each term on both the right and the left sides of the equation by -1 so that a is not negative.

1. Write the equation that represents the line shown.

   Use the properties of equality to change the equation from slope-intercept form, $y = mx + b$, to standard form, $ax + by = c$, where a, b, and c are integers, and a is not negative.

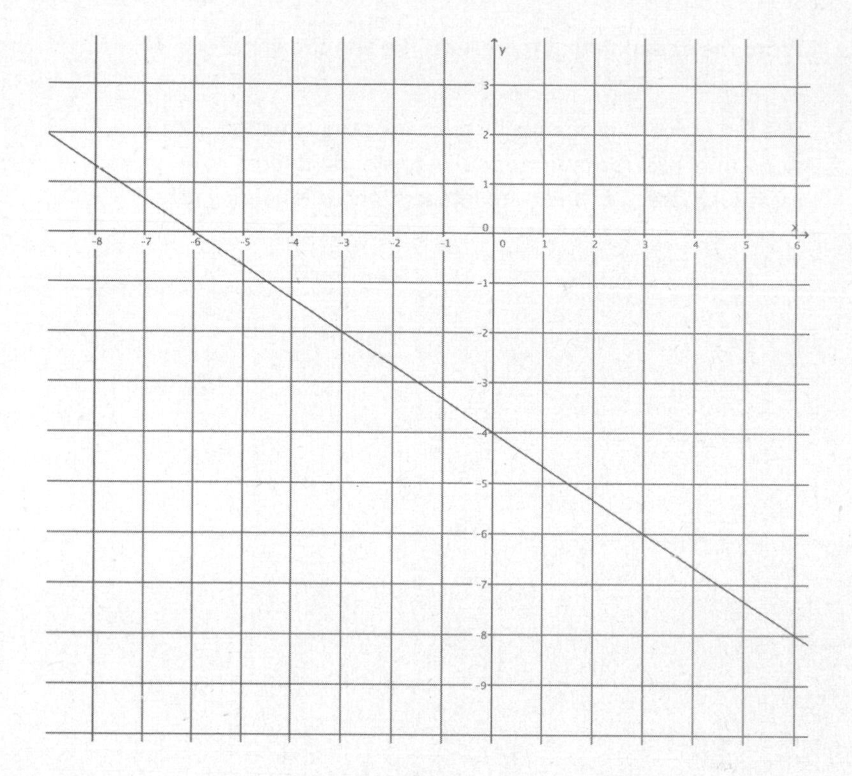

2. Write the equation that represents the line shown.

   Use the properties of equality to change the equation from slope-intercept form, $y = mx + b$, to standard form, $ax + by = c$, where a, b, and c are integers, and a is not negative.

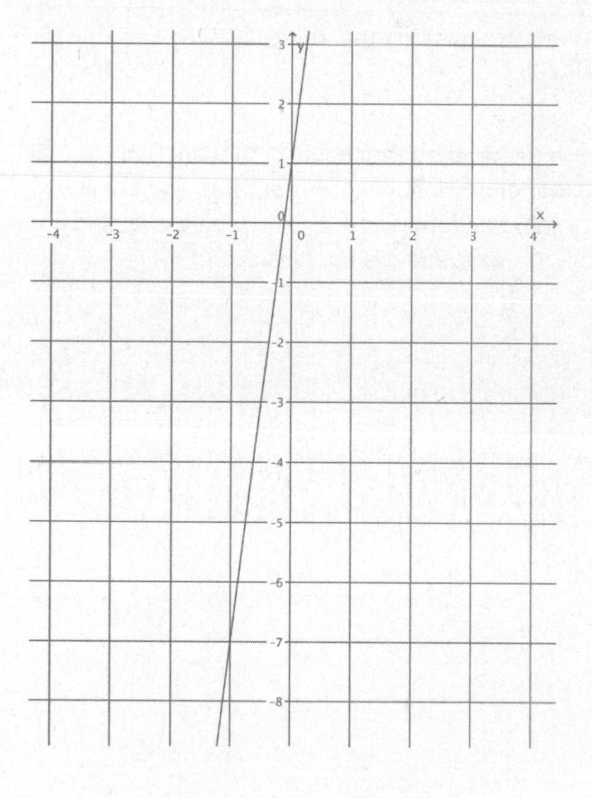

3. Write the equation that represents the line shown.

   Use the properties of equality to change the equation from slope-intercept form, $y = mx + b$, to standard form, $ax + by = c$, where a, b, and c are integers, and a is not negative.

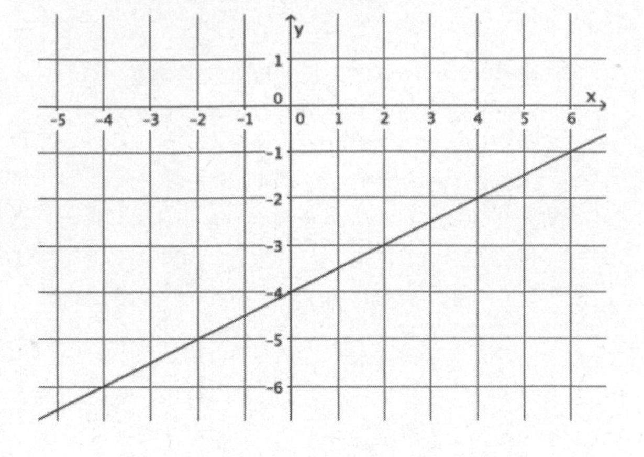

4. Write the equation that represents the line shown.

   Use the properties of equality to change the equation from slope-intercept form, $y = mx + b$, to standard form, $ax + by = c$, where a, b, and c are integers, and a is not negative.

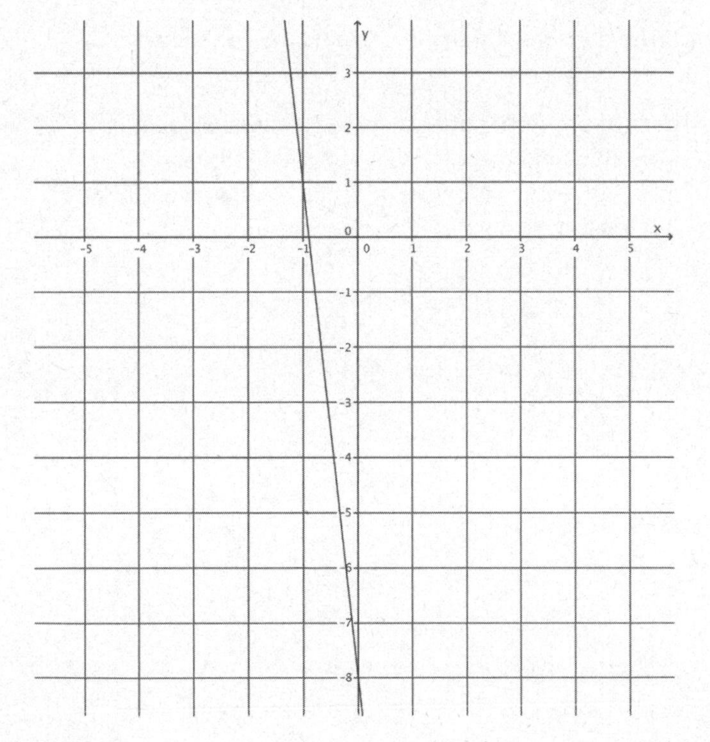

5. Write the equation that represents the line shown.

   Use the properties of equality to change the equation from slope-intercept form, $y = mx + b$, to standard form, $ax + by = c$, where a, b, and c are integers, and a is not negative.

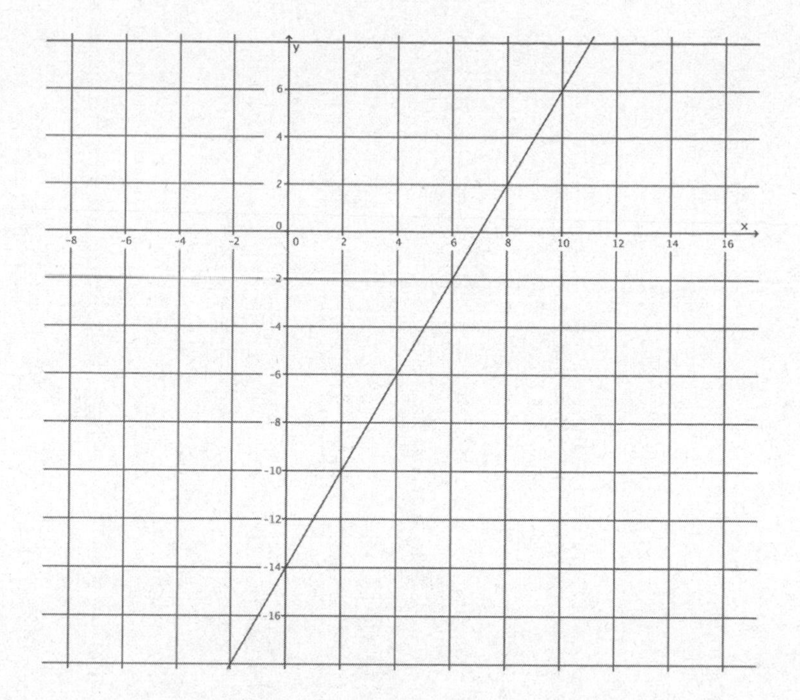

6. Write the equation that represents the line shown.

   Use the properties of equality to change the equation from slope-intercept form, $y = mx + b$, to standard form, $ax + by = c$, where a, b, and c are integers, and a is not negative.

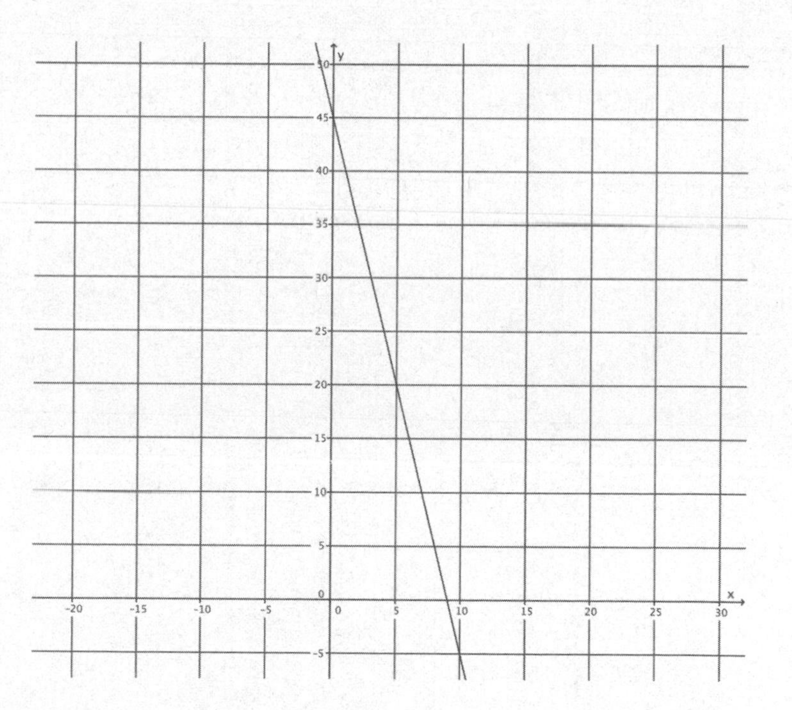

Example 1

Let a line l be given in the coordinate plane. What linear equation is the graph of line l?

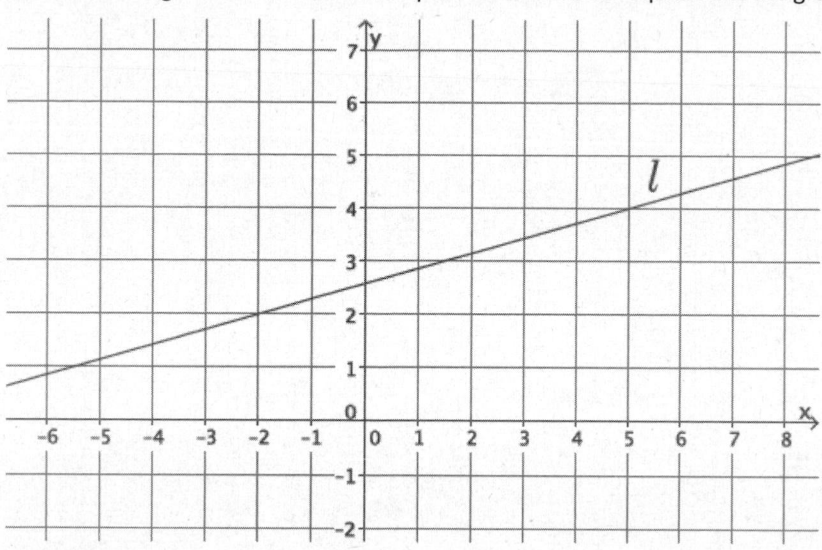

Example 2

Let a line l be given in the coordinate plane. What linear equation is the graph of line l?

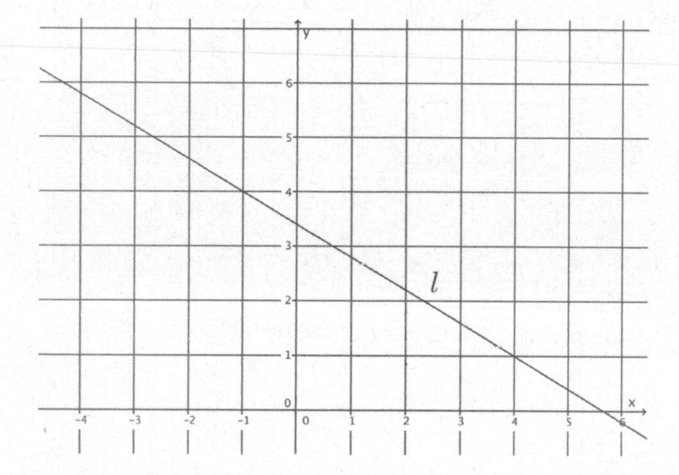

Example 3

Let a line l be given in the coordinate plane. What linear equation is the graph of line l?

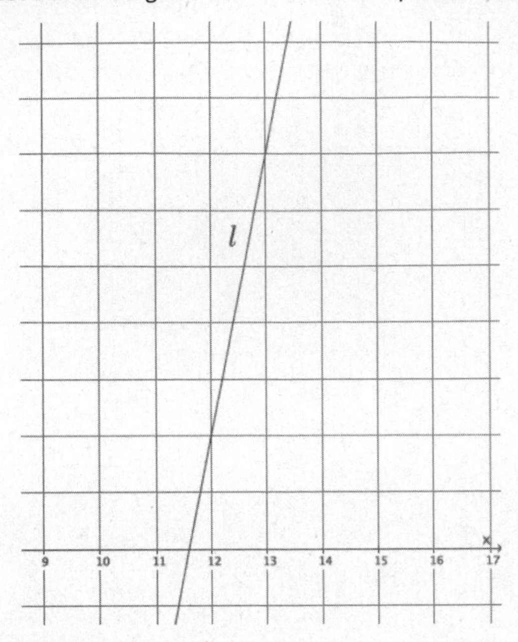

Example 4

Let a line l be given in the coordinate plane. What linear equation is the graph of line l?

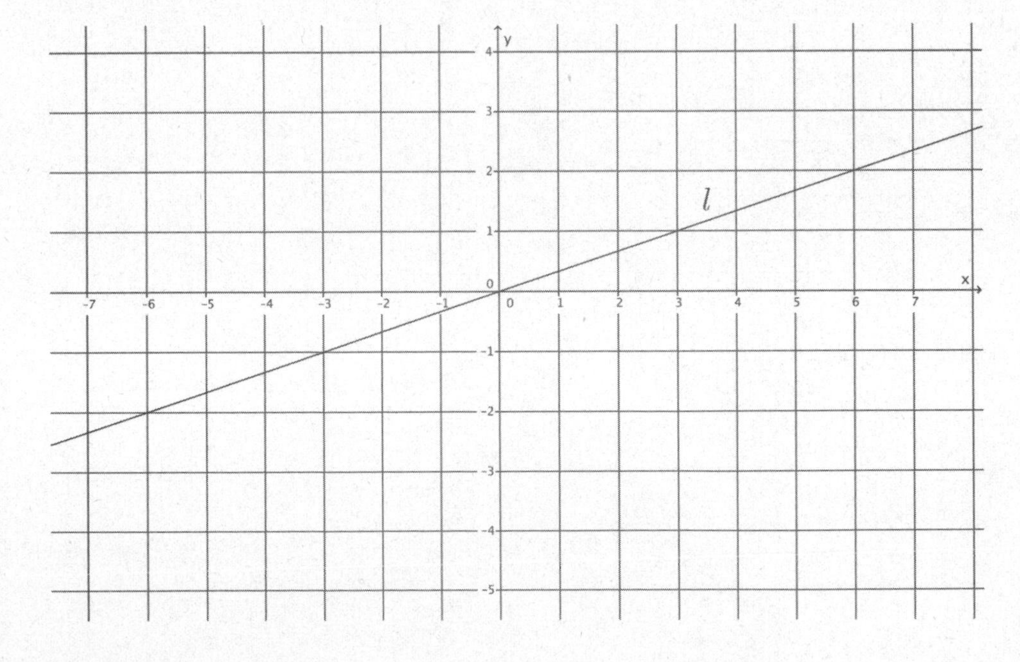

Exercises

1. Write the equation for the line l shown in the figure.

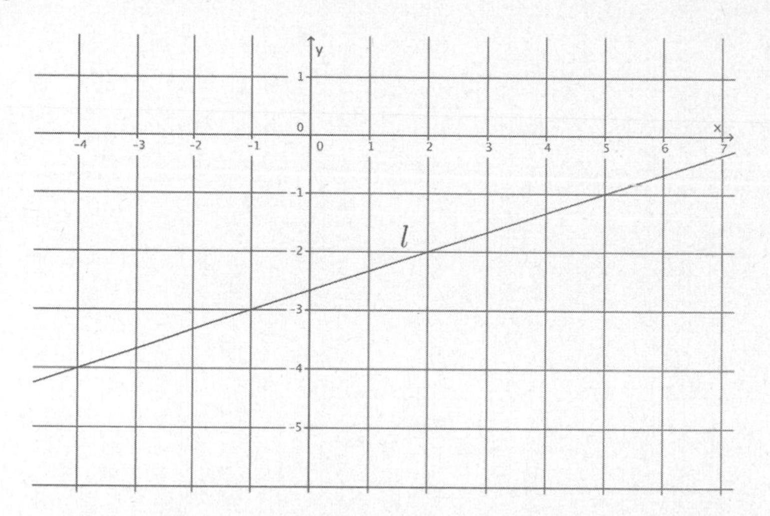

2. Write the equation for the line l shown in the figure.

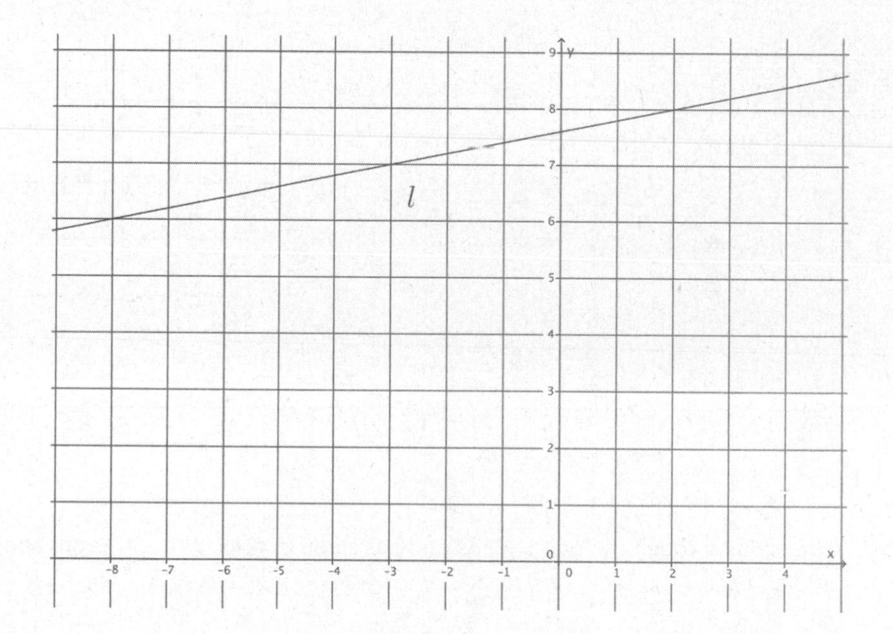

3. Determine the equation of the line that goes through points $(-4, 5)$ and $(2, 3)$.

4. Write the equation for the line l shown in the figure.

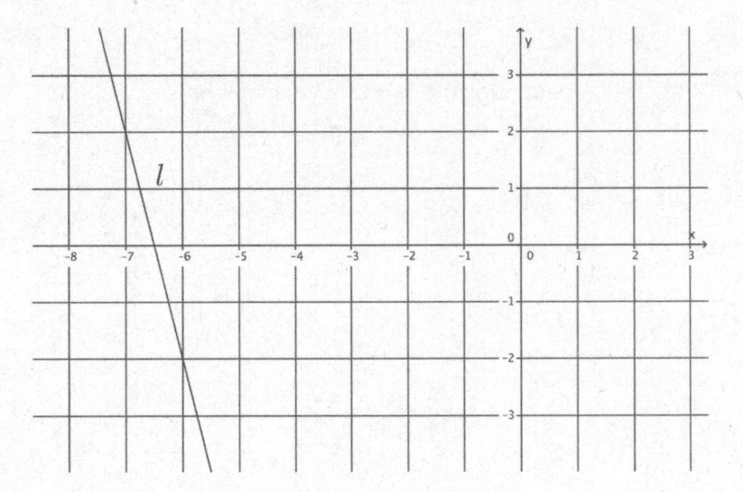

5. A line goes through the point $(8, 3)$ and has slope $m = 4$. Write the equation that represents the line.

Lesson Summary

Let (x_1, y_1) and (x_2, y_2) be the coordinates of two distinct points on a non-vertical line in a coordinate plane. We find the slope of the line by

$$m = \frac{y_2 - y_1}{x_2 - x_1}.$$

This version of the slope formula, using coordinates of x and y instead of p and r, is a commonly accepted version.

As soon as you multiply the slope by the denominator of the fraction above, you get the following equation:

$$m(x_2 - x_1) = y_2 - y_1.$$

This form of an equation is referred to as the *point-slope form* of a linear equation.

Given a known (x, y), then the equation is written as

$$m(x - x_1) = (y - y_1).$$

The following is slope-intercept form of a line:

$$y = mx + b.$$

In this equation, m is slope, and $(0, b)$ is the y-intercept point.

To write the equation of a line, you must have two points, one point and slope, or a graph of the line.

Name ______________________________ Date ______________

1. Write the equation for the line l shown in the figure below.

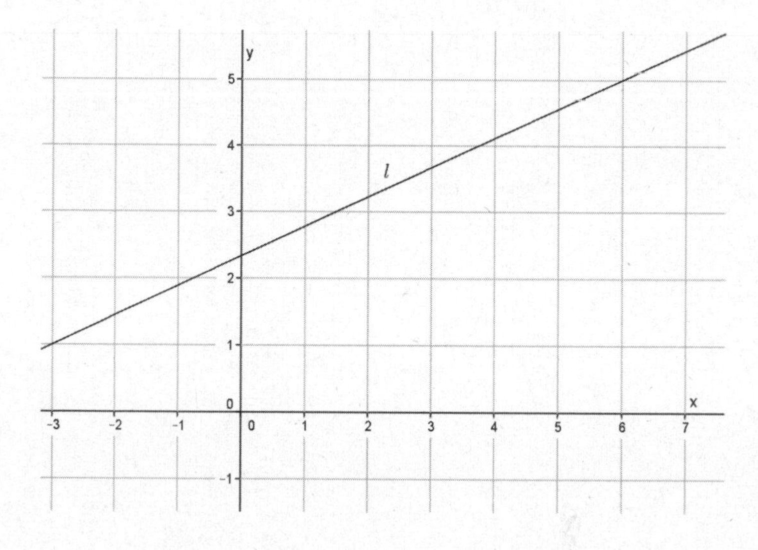

2. A line goes through the point $(5, -7)$ and has slope $m = -3$. Write the equation that represents the line.

1. Write the equation for the line l shown in the figure.

> I need to identify two points to find the slope. I will use $(-3, -2)$ and $(4, 4)$ because they have integer coordinates.

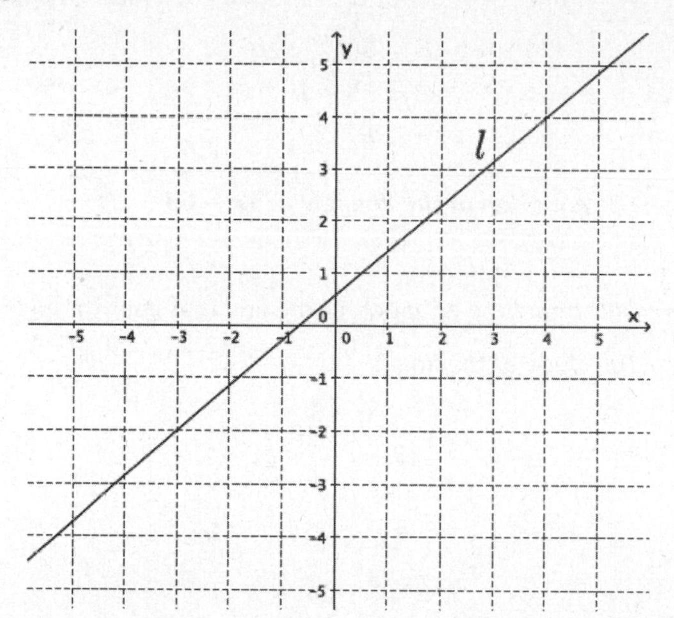

Using the points $(-3, -2)$ and $(4, 4)$, the slope of the line is

$$m = \frac{4 - (-2)}{4 - (-3)}$$

$$= \frac{6}{7}$$

The y-intercept point of the line is

$$4 = \frac{6}{7}(4) + b$$

$$4 = \frac{24}{7} + b$$

$$4 - \frac{24}{7} = \frac{24}{7} - \frac{24}{7} + b$$

$$\frac{4}{7} = b$$

> I can see that the line doesn't intersect the y-axis at integer coordinates, so I need to calculate the y-intercept, $(0, b)$. I can use either point to substitute into my equation $y = mx + b$.

The equation of the line is $y = \frac{6}{7}x + \frac{4}{7}$.

2. Write the equation for the line that goes through point $(11, -8)$ with slope $m = 5$.

$$\mathbf{-8 = 5(11) + b}$$
$$\mathbf{-8 = 55 + b}$$
$$\mathbf{-63 = b}$$

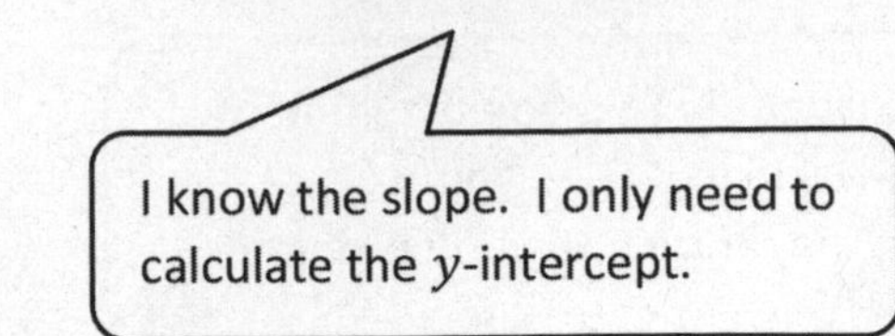

The equation of the line is $\boldsymbol{y = 5x - 63}$.

3. Determine the equation of the line that goes through points $(-7,3)$ and $(5,-6)$.

The slope of the line is

$$\mathbf{m = \frac{-6 - 3}{5 - (-7)} = \frac{-9}{12} = -\frac{3}{4}}$$

This problem is similar to Problem 1, but without a graph.

The y-intercept point of the line is

$$\mathbf{-6 = -\frac{3}{4}(5) + b}$$
$$\mathbf{-6 = -\frac{15}{4} + b}$$
$$\mathbf{-\frac{9}{4} = b}$$

The equation of the line is $\boldsymbol{y = -\frac{3}{4}x - \frac{9}{4}}$.

1. Write the equation for the line l shown in the figure.

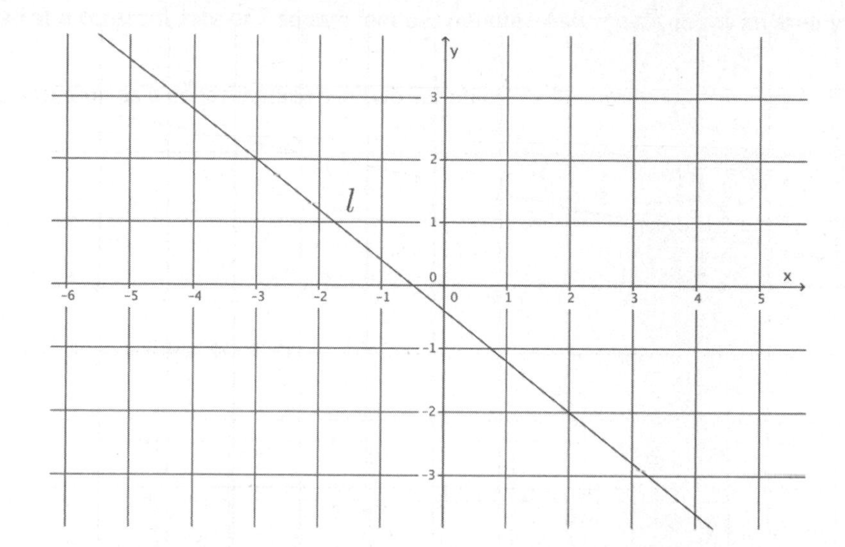

2. Write the equation for the line l shown in the figure.

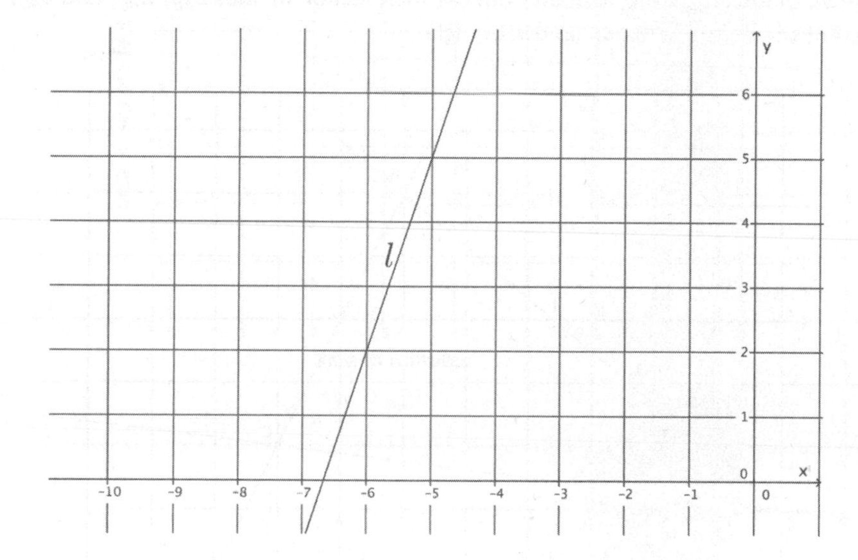

c. Using the graph or the equation, determine the total area he paints after 8 minutes, $1\frac{1}{2}$ hours, and 2 hours. Note that the units are in minutes and hours.

2. The figure below represents Nathan's constant rate of walking.

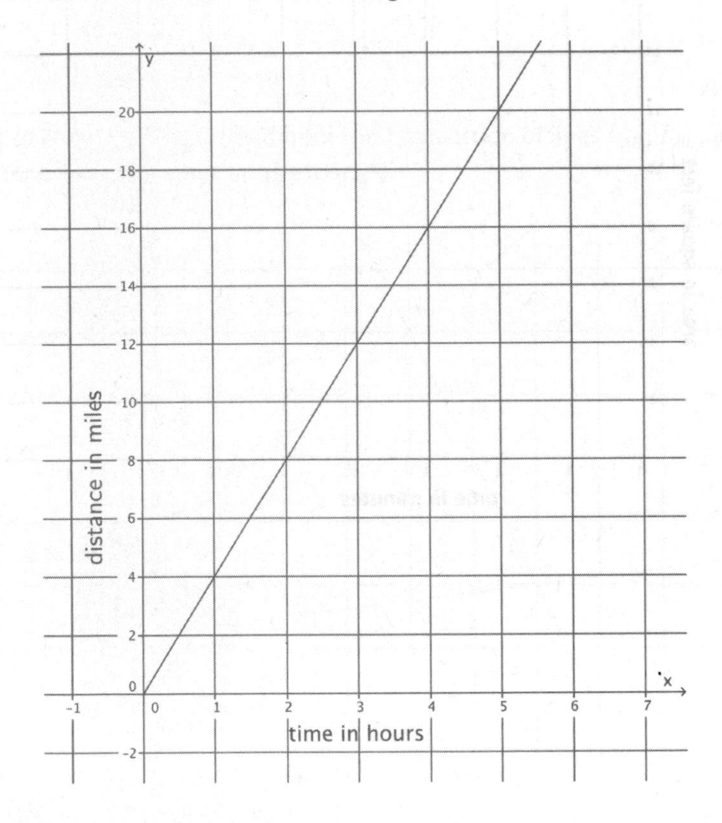

a. Nicole just finished a 5-mile walkathon. It took her 1.4 hours. Assume she walks at a constant rate. Let y represent the distance Nicole walks in x hours. Describe Nicole's walking at a constant rate as a linear equation in two variables.

b. Who walks at a greater speed? Explain.

3.

a. Susan can type 4 pages of text in 10 minutes. Assuming she types at a constant rate, write the linear equation that represents the situation.

b. The table of values below represents the number of pages that Anne can type, y, in a few selected x minutes. Assume she types at a constant rate.

Minutes (x)	**Pages Typed** (y)
3	2
5	$\frac{10}{3}$
8	$\frac{16}{3}$
10	$\frac{20}{3}$

Who types faster? Explain.

4.

a. Phil can build 3 birdhouses in 5 days. Assuming he builds birdhouses at a constant rate, write the linear equation that represents the situation.

b. The figure represents Karl's constant rate of building the same kind of birdhouses. Who builds birdhouses faster? Explain.

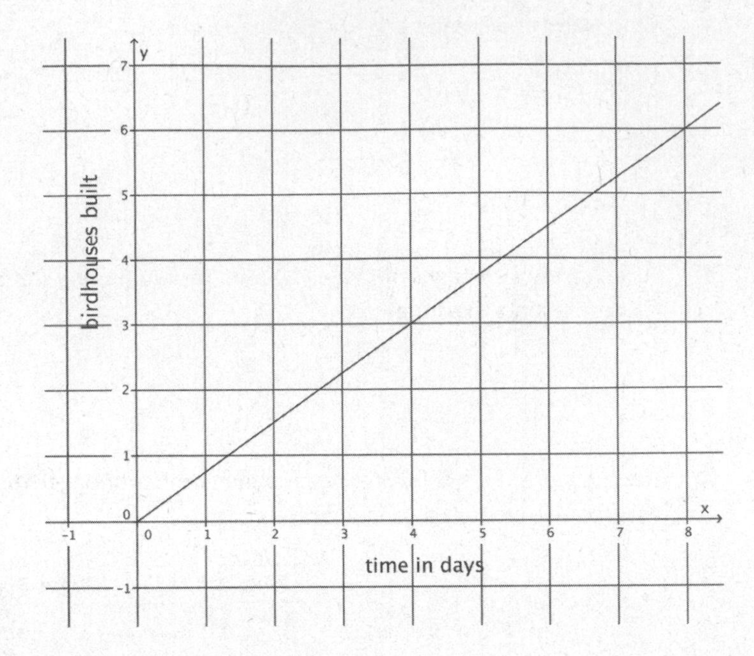

5. Explain your general strategy for comparing proportional relationships.

Lesson Summary

Problems involving constant rate can be expressed as linear equations in two variables.

When given information about two proportional relationships, their rates of change can be compared by comparing the slopes of the graphs of the two proportional relationships.

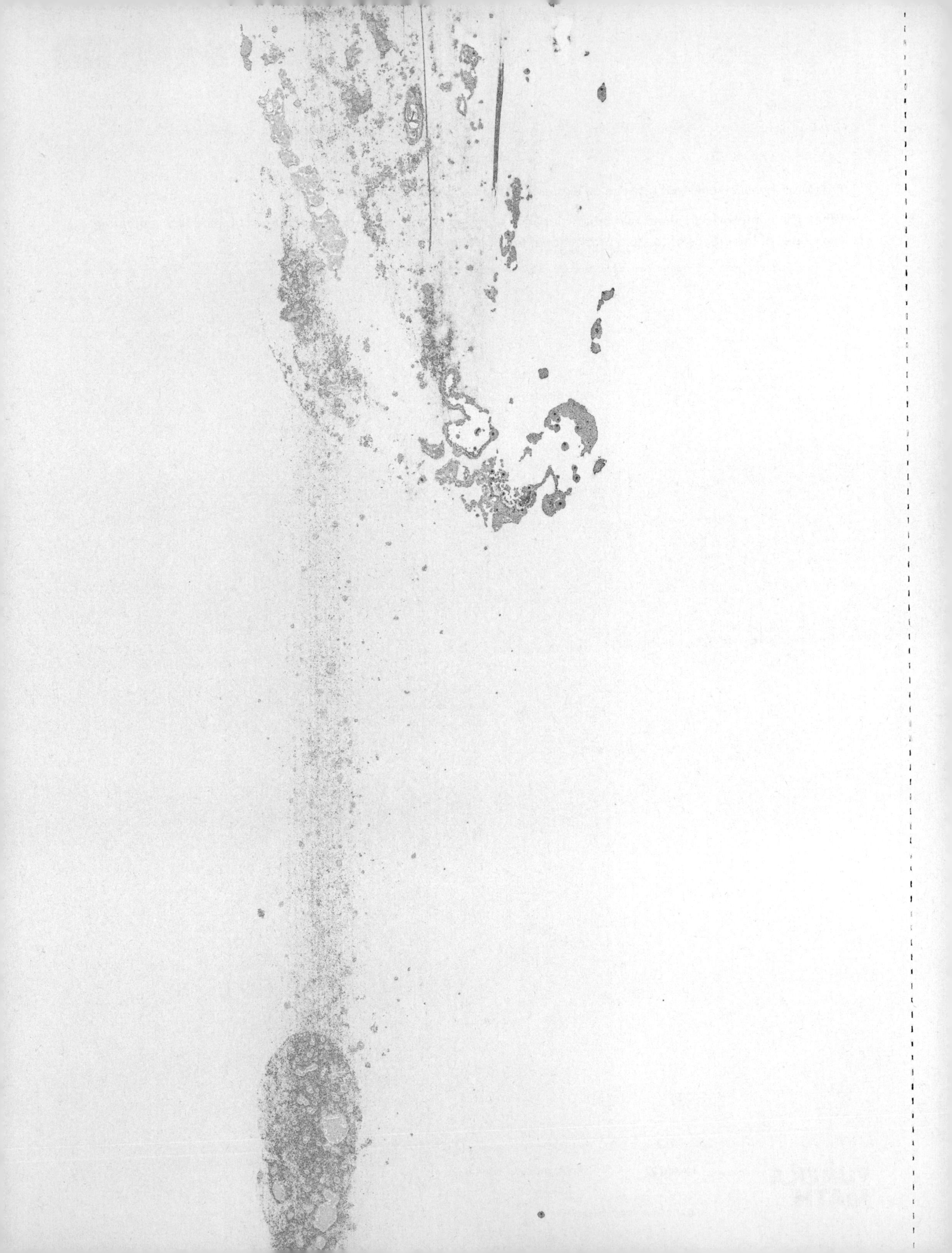

Name ______________________________ Date ______________

1. Water flows out of Pipe A at a constant rate. Pipe A can fill 3 buckets of the same size in 14 minutes. Write a linear equation that represents the situation.

2. The figure below represents the rate at which Pipe B can fill the same-sized buckets.

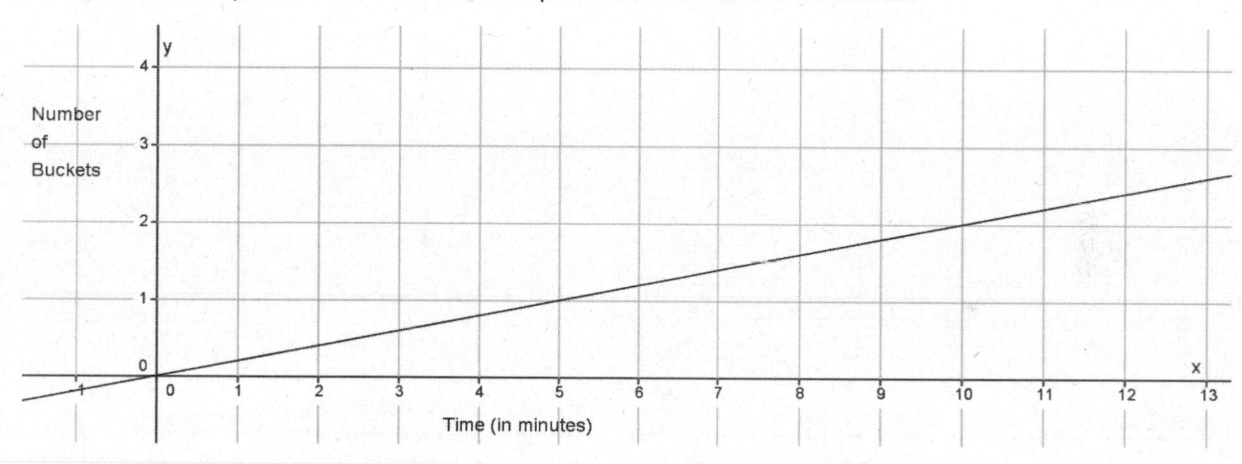

Which pipe fills buckets faster? Explain.

1. Train A can travel a distance of 450 miles in 7 hours.

   a. Assuming the train travels at a constant rate, write the linear equation that represents the situation.

   Let y represent the total number of miles Train A travels in x hours. We can write $\frac{y}{x} = \frac{450}{7}$ and $y = \frac{450}{7}x$.

   b. The figure represents the constant rate of travel for Train B. Which train is faster? Explain.

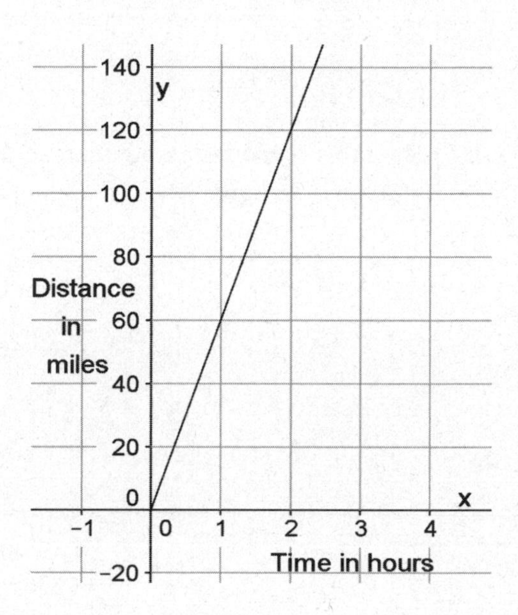

To see which train is faster, I need to compare the slopes or rates of change.

Train A is faster than Train B. The slope, or rate, for Train A is $\frac{450}{7}$, and the slope of the line for Train B is $\frac{60}{1}$. When you compare the slopes, you see that $\frac{450}{7} > 60$.

2. Norton and Sylvia read the same book. Norton can read 33 pages in 8 minutes.

 a. Assuming he reads at a constant rate, write the linear equation that represents the situation.

 Let y represent the total number of pages Norton can read in x minutes. We can write $\frac{y}{x} = \frac{33}{8}$ and $y = \frac{33}{8}x$.

 b. The table of values below represents the number of pages read by Sylvia for a few selected time intervals. Assume Sylvia is reading at a constant rate. Who reads faster? Explain.

Minutes (x)	Pages Read (y)
3	11
5	$\frac{55}{3}$
6	22
8	$\frac{88}{3}$

Since Sylvia is reading at a constant rate, I can use any two points to calculate the slope or rate of change.

 Norton reads faster. Using the table of values, I can find the slope that represents Sylvia's constant rate of reading: $\frac{11}{3}$. The slope or rate for Norton is $\frac{33}{8}$. When you compare the slopes, you see that $\frac{33}{8} > \frac{11}{3}$.

1.

a. Train A can travel a distance of 500 miles in 8 hours. Assuming the train travels at a constant rate, write the linear equation that represents the situation.

b. The figure represents the constant rate of travel for Train B.

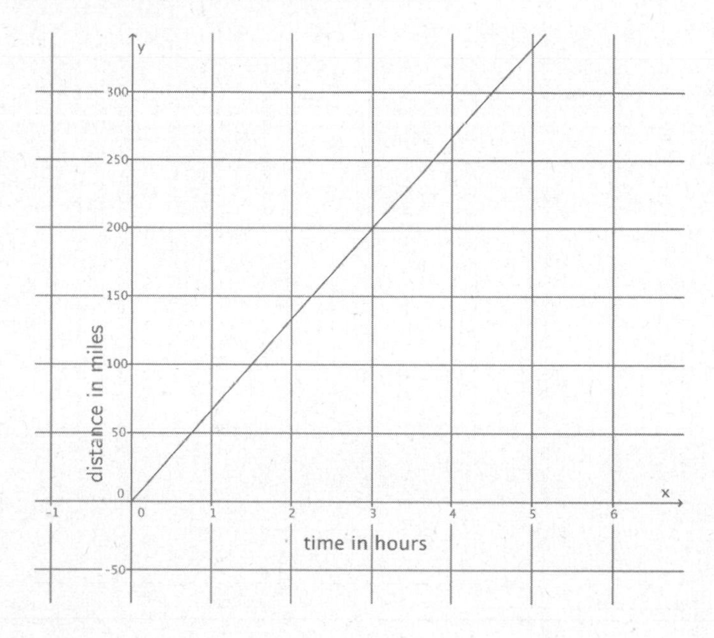

Which train is faster? Explain.

2.

a. Natalie can paint 40 square feet in 9 minutes. Assuming she paints at a constant rate, write the linear equation that represents the situation.

b. The table of values below represents the area painted by Steven for a few selected time intervals. Assume Steven is painting at a constant rate.

Minutes (x)	Area Painted (y)
3	10
5	$\frac{50}{3}$
6	20
8	$\frac{80}{3}$

Who paints faster? Explain.

3.

a. Bianca can run 5 miles in 41 minutes. Assuming she runs at a constant rate, write the linear equation that represents the situation.

b. The figure below represents Cynthia's constant rate of running.

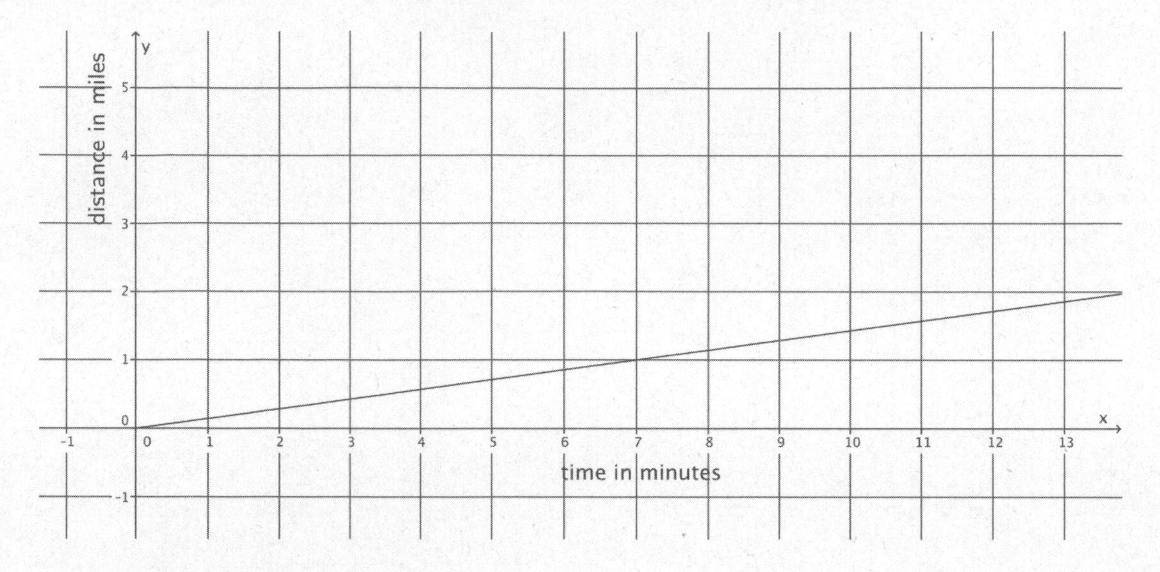

Who runs faster? Explain.

4.

a. Geoff can mow an entire lawn of 450 square feet in 30 minutes. Assuming he mows at a constant rate, write the linear equation that represents the situation.

b. The figure represents Mark's constant rate of mowing a lawn.

Who mows faster? Explain.

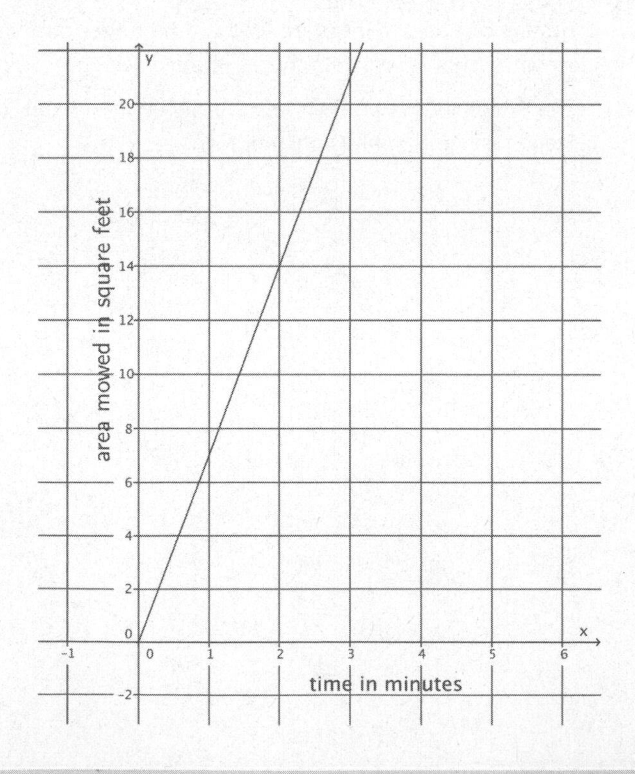

5.

a. Juan can walk to school, a distance of 0.75 mile, in 8 minutes. Assuming he walks at a constant rate, write the linear equation that represents the situation.

b. The figure below represents Lena's constant rate of walking.

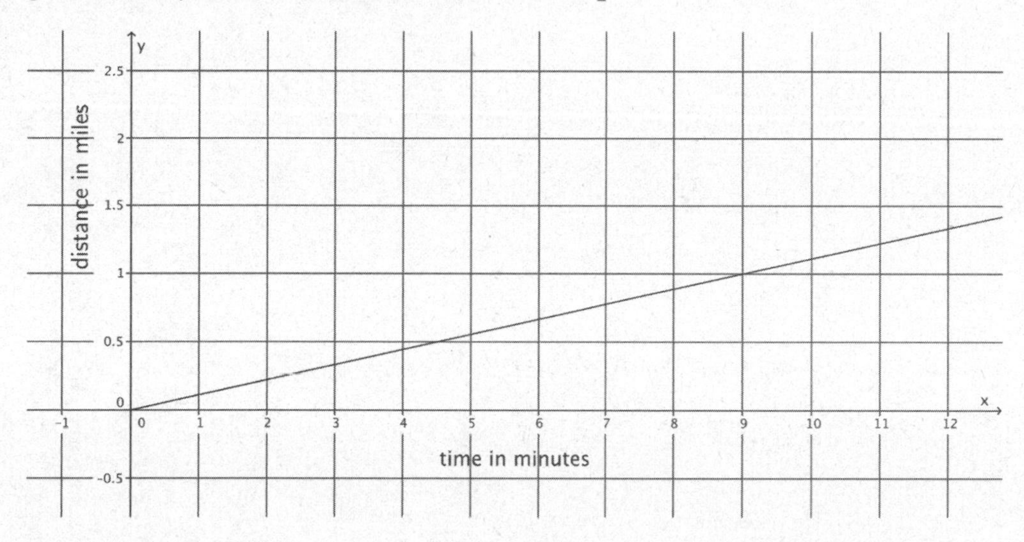

Who walks faster? Explain.

Exploratory Challenge/Exercises 1–3

1. Sketch the graph of the equation $9x + 3y = 18$ using intercepts. Then, answer parts (a)–(f) that follow.

   a. Sketch the graph of the equation $y = -3x + 6$ on the same coordinate plane.

   b. What do you notice about the graphs of $9x + 3y = 18$ and $y = -3x + 6$? Why do you think this is so?

   c. Rewrite $y = -3x + 6$ in standard form.

   d. Identify the constants a, b, and c of the equation in standard form from part (c).

e. Identify the constants of the equation $9x + 3y = 18$. Note them as a', b', and c'.

f. What do you notice about $\frac{a'}{a}$, $\frac{b'}{b}$, and $\frac{c'}{c}$?

2. Sketch the graph of the equation $y = \frac{1}{2}x + 3$ using the y-intercept point and the slope. Then, answer parts (a)–(f) that follow.

a. Sketch the graph of the equation $4x - 8y = -24$ using intercepts on the same coordinate plane.

b. What do you notice about the graphs of $y = \frac{1}{2}x + 3$ and $4x - 8y = -24$? Why do you think this is so?

c. Rewrite $y = \frac{1}{2}x + 3$ in standard form.

d. Identify the constants a, b, and c of the equation in standard form from part (c).

e. Identify the constants of the equation $4x - 8y = -24$. Note them as a', b', and c'.

f. What do you notice about $\frac{a'}{a}$, $\frac{b'}{b}$, and $\frac{c'}{c}$?

3. The graphs of the equations $y = \frac{2}{3}x - 4$ and $6x - 9y = 36$ are the same line.

a. Rewrite $y = \frac{2}{3}x - 4$ in standard form.

b. Identify the constants a, b, and c of the equation in standard form from part (a).

c. Identify the constants of the equation $6x - 9y = 36$. Note them as a', b', and c'.

d. What do you notice about $\frac{a'}{a}$, $\frac{b'}{b}$, and $\frac{c'}{c}$?

e. You should have noticed that each fraction was equal to the same constant. Multiply that constant by the standard form of the equation from part (a). What do you notice?

Exercises 4–8

4. Write three equations whose graphs are the same line as the equation $3x + 2y = 7$.

5. Write three equations whose graphs are the same line as the equation $x - 9y = \frac{3}{4}$.

6. Write three equations whose graphs are the same line as the equation $-9x + 5y = -4$.

7. Write at least two equations in the form $ax + by = c$ whose graphs are the line shown below.

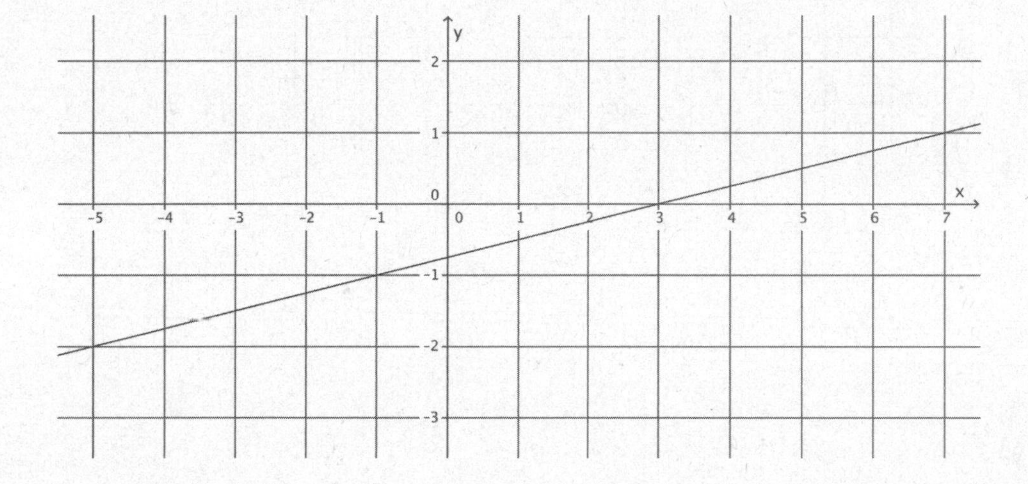

8. Write at least two equations in the form $ax + by = c$ whose graphs are the line shown below.

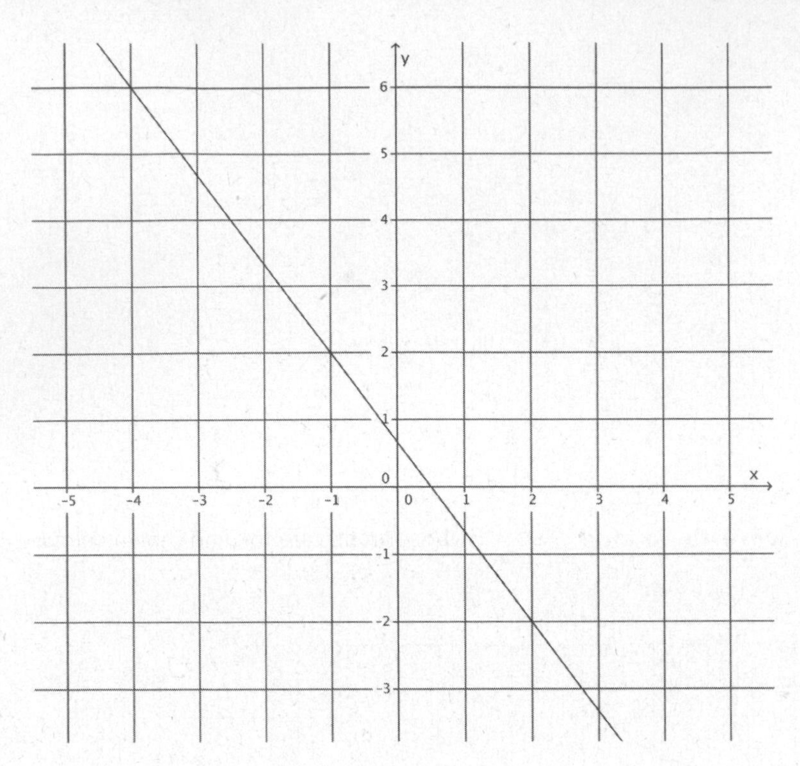

Lesson Summary

Two equations define the same line if the graphs of those two equations are the same given line. Two equations that define the same line are the same equation, just in different forms. The equations may look different (different constants, different coefficients, or different forms).

When two equations are written in standard form, $ax + by = c$ and $a'x + b'y = c'$, they define the same line when $\frac{a'}{a} = \frac{b'}{b} = \frac{c'}{c}$ is true.

Name ____ Date ____________________

1. Do the graphs of the equations $-16x + 12y = 33$ and $-4x + 3y = 8$ graph as the same line? Why or why not?

2. Given the equation $3x - y = 11$, write another equation that will have the same graph. Explain why.

1. Do the equations $3x - 5y = 8$ and $6x - 10y = 16$ define the same line? Explain.

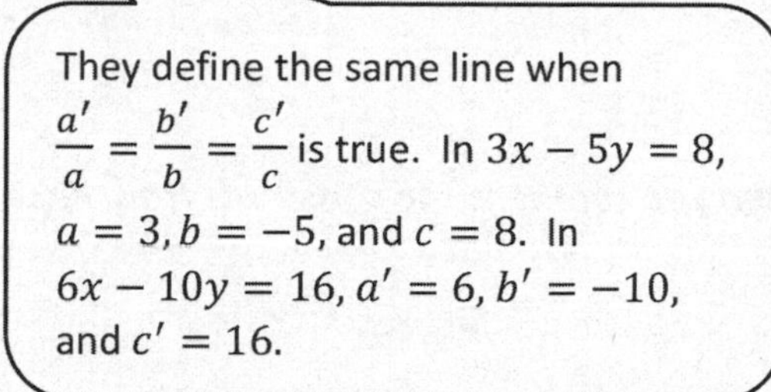

Yes, these equations define the same line. When you compare the constants from each equation, you get

$$\frac{a'}{a} = \frac{6}{3} = 2, \frac{b'}{b} = \frac{-10}{-5} = 2, \textit{ and } \frac{c'}{c} = \frac{16}{8} = 2.$$

When I multiply the first equation by 2, I get the second equation.

$$(3x - 5y = 8)2$$
$$6x - 10y = 16$$

Therefore, these equations define the same line.

2. Do the equations $y = -\frac{7}{5}x - 4$ and $14x + 10y = -40$ define the same line? Explain.

I need to rewrite the first equation in standard form before I can determine if they define the same line.

Yes, these equations define the same line. When you rewrite the first equation in standard form:

$$y = -\frac{7}{5}x - 4$$
$$\left(y = -\frac{7}{5}x - 4\right)5$$
$$5y = -7x - 20$$
$$7x + 5y = -20$$

When you compare the constants from each equation:

$$\frac{a'}{a} = \frac{14}{7} = 2, \frac{b'}{b} = \frac{10}{5} = 2, \textit{ and } \frac{c'}{c} = \frac{-40}{-20} = 2.$$

When I multiply the first equation by 2, I get the second equation.

$$(7x + 5y = -20)2$$
$$14x + 10y = -40$$

Therefore, these equations define the same line.

3. Write an equation that would define the same line as $9x - 12y = 15$.

Answers will vary. When you multiply the equation by 2:

$$(9x - 12y = 15)2$$
$$18x - 24y = 30$$

When you compare the constants from each equation:

$$\frac{a'}{a} = \frac{18}{9} = 2,\ \frac{b'}{b} = \frac{-24}{-12} = 2,\ and\ \frac{c'}{c} = \frac{30}{15} = 2.$$

Therefore, these equations define the same line.

I can multiply the equation by any number other than zero and then make sure that $\frac{a'}{a}, \frac{b'}{b}, \frac{c'}{c}$ are all equal to the same number.

4. Challenge: Show that if the two lines given by $ax + by = c$ and $a'x + b'y = c'$ are the same when $b = 0$ (vertical lines), then there exists a nonzero number s so that $a' = sa$, $b' = sb$, and $c' = sc$.

When $b = 0$, then $b' = 0$, and the equations are $ax = c$ and $a'x = c'$.

I need to write the equations when $b = 0$. Since the problem said they were the same line, I will solve for x in both equations so that I can use substitution.

We can rewrite the equations as $x = \frac{c}{a}$ and $x = \frac{c'}{a'}$. Because the equations graph as the same line, then we know that

$$\frac{c}{a} = \frac{c'}{a'}$$

and we can rewrite those fractions as

$$\frac{a'}{a} = \frac{c'}{c}.$$

I can use properties of equality to rewrite in the form I need.

These fractions are equal to the same number. Let that number be s. Then $\frac{a'}{a} = s$ and $\frac{c'}{c} = s$. Therefore, $a' = sa$ and $c' = sc$.

1. Do the equations $x + y = -2$ and $3x + 3y = -6$ define the same line? Explain.

2. Do the equations $y = -\frac{5}{4}x + 2$ and $10x + 8y = 16$ define the same line? Explain.

3. Write an equation that would define the same line as $7x - 2y = 5$.

4. Challenge: Show that if the two lines given by $ax + by = c$ and $a'x + b'y = c'$ are the same when $b = 0$ (vertical lines), then there exists a nonzero number s so that $a' = sa$, $b' = sb$, and $c' = sc$.

5. Challenge: Show that if the two lines given by $ax + by = c$ and $a'x + b'y = c'$ are the same when $a = 0$ (horizontal lines), then there exists a nonzero number s so that $a' = sa$, $b' = sb$, and $c' = sc$.

Exercises

1. Derek scored 30 points in the basketball game he played, and not once did he go to the free throw line. That means that Derek scored two-point shots and three-point shots. List as many combinations of two- and three-pointers as you can that would total 30 points.

Number of Two-Pointers	Number of Three-Pointers

Write an equation to describe the data.

2. Derek tells you that the number of two-point shots that he made is five more than the number of three-point shots. How many combinations can you come up with that fit this scenario? (Don't worry about the total number of points.)

Number of Two-Pointers	Number of Three-Pointers

Write an equation to describe the data.

3. Which pair of numbers from your table in Exercise 2 would show Derek's actual score of 30 points?

4. Efrain and Fernie are on a road trip. Each of them drives at a constant speed. Efrain is a safe driver and travels 45 miles per hour for the entire trip. Fernie is not such a safe driver. He drives 70 miles per hour throughout the trip. Fernie and Efrain left from the same location, but Efrain left at 8:00 a.m., and Fernie left at 11:00 a.m. Assuming they take the same route, will Fernie ever catch up to Efrain? If so, approximately when?

a. Write the linear equation that represents Efrain's constant speed. Make sure to include in your equation the extra time that Efrain was able to travel.

b. Write the linear equation that represents Fernie's constant speed.

c. Write the system of linear equations that represents this situation.

d. Sketch the graphs of the two linear equations.

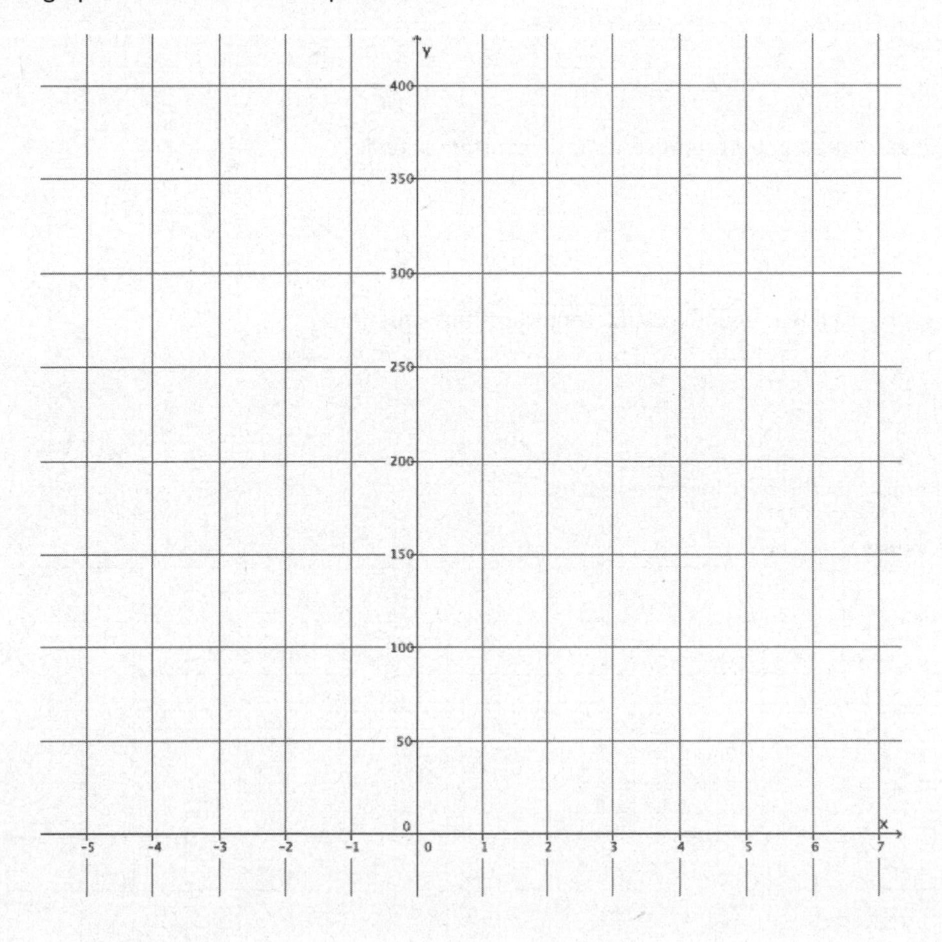

e. Will Fernie ever catch up to Efrain? If so, approximately when?

f. At approximately what point do the graphs of the lines intersect?

5. Jessica and Karl run at constant speeds. Jessica can run 3 miles in 24 minutes. Karl can run 2 miles in 14 minutes. They decide to race each other. As soon as the race begins, Karl trips and takes 2 minutes to recover.

   a. Write the linear equation that represents Jessica's constant speed. Make sure to include in your equation the extra time that Jessica was able to run.

   b. Write the linear equation that represents Karl's constant speed.

   c. Write the system of linear equations that represents this situation.

   d. Sketch the graphs of the two linear equations.

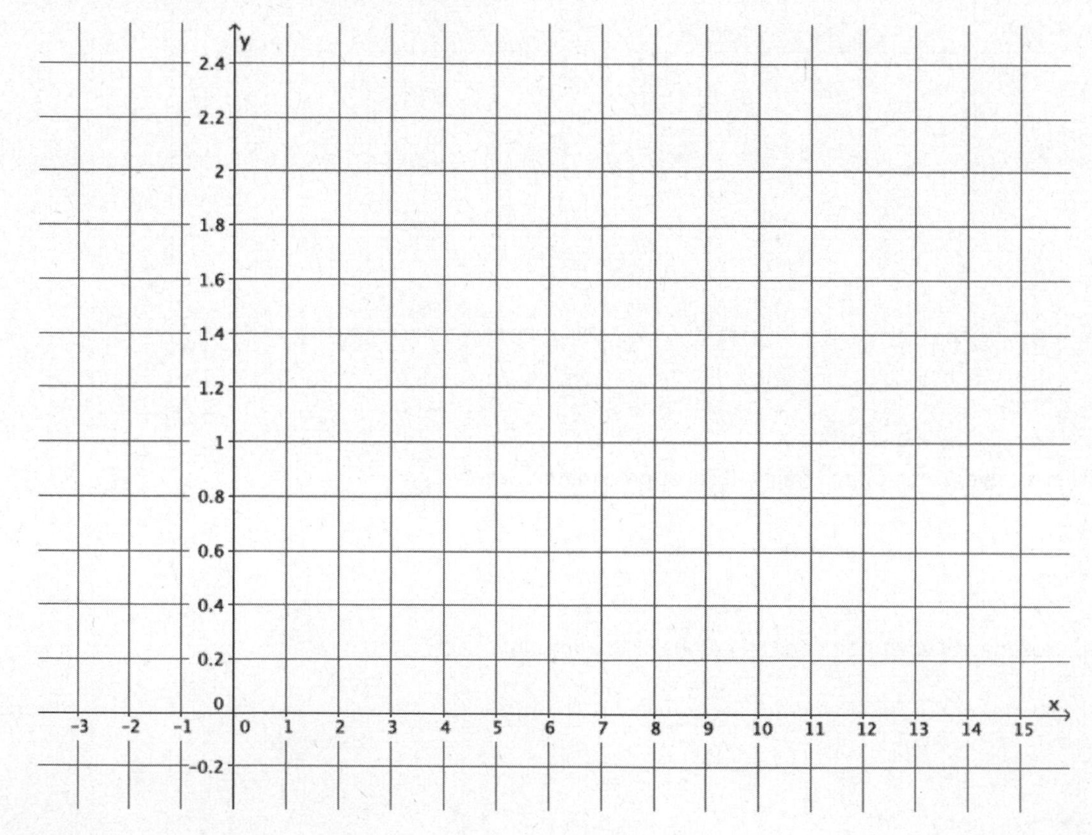

e. Use the graph to answer the questions below.

i. If Jessica and Karl raced for 3 miles, who would win? Explain.

ii. At approximately what point would Jessica and Karl be tied? Explain.

Lesson Summary

A *system of linear equations* is a set of two or more linear equations. When graphing a pair of linear equations in two variables, both equations in the system are graphed on the same coordinate plane.

A *solution to a system of two linear equations in two variables* is an ordered pair of numbers that is a solution to both equations. For example, the solution to the system of linear equations $\begin{cases} x + y = 6 \\ x - y = 4 \end{cases}$ is the ordered pair $(5, 1)$ because substituting 5 in for x and 1 in for y results in two true equations: $5 + 1 = 6$ and $5 - 1 = 4$.

Systems of linear equations are notated using brackets to group the equations, for example: $\begin{cases} y = \frac{1}{8}x + \frac{5}{2} \\ y = \frac{4}{25}x \end{cases}$.

Name____________________________________ Date ____________________

Darnell and Hector ride their bikes at constant speeds. Darnell leaves Hector's house to bike home. He can bike the 8 miles in 32 minutes. Five minutes after Darnell leaves, Hector realizes that Darnell left his phone. Hector rides to catch up. He can ride to Darnell's house in 24 minutes. Assuming they bike the same path, will Hector catch up to Darnell before he gets home?

a. Write the linear equation that represents Darnell's constant speed.

b. Write the linear equation that represents Hector's constant speed. Make sure to take into account that Hector left after Darnell.

c. Write the system of linear equations that represents this situation.

d. Sketch the graphs of the two equations.

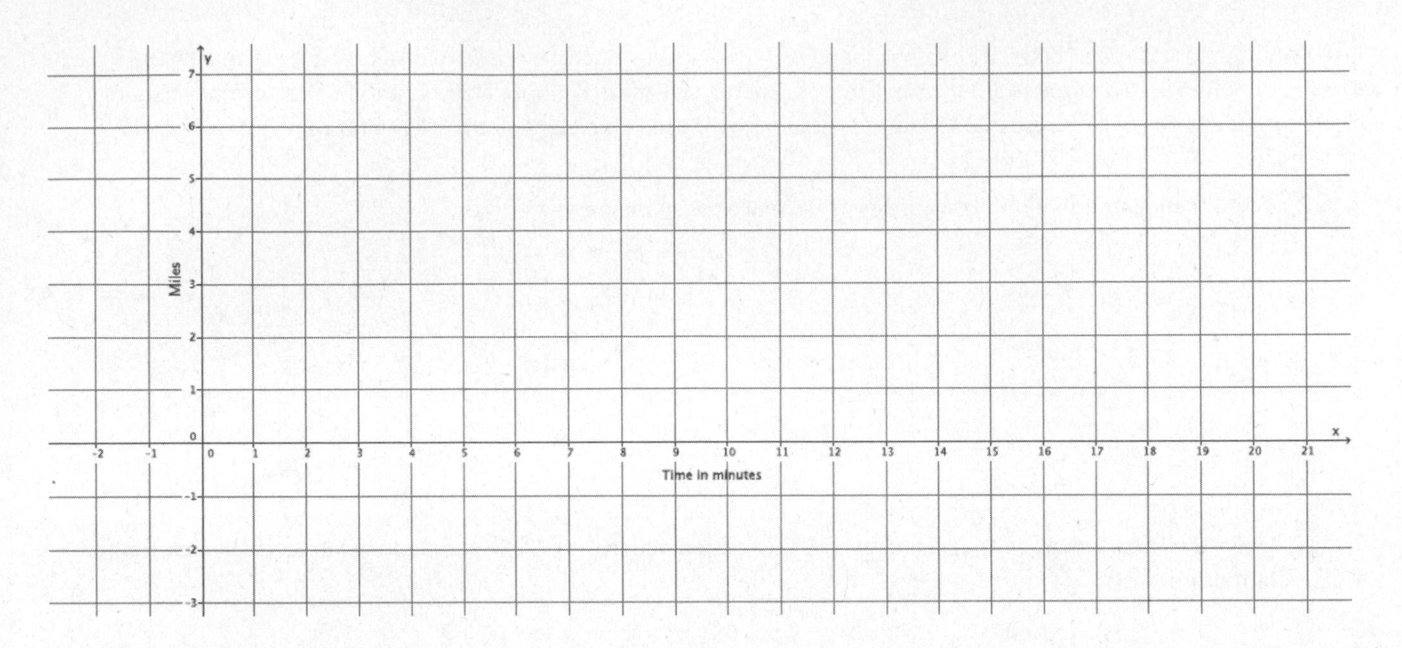

e. Will Hector catch up to Darnell before he gets home? If so, approximately when?

f. At approximately what point do the graphs of the lines intersect?

1. Allen and Regina walk at constant speeds. Allen can walk 1 mile in 60 minutes, and Regina can walk 2 miles in 90 minutes. Regina started walking 10 minutes after Allen. Assuming they walk the same path, when will Regina catch up to Allen?

   a. Write the linear equation that represents Regina's constant speed.

   Regina's rate is $\frac{2}{90}$ miles per minute, which is the same as $\frac{1}{45}$ miles per minute. If Regina continues walking y miles in x minutes at a constant speed, then $y = \frac{1}{45}x$

   Since they are walking at constant speeds, I can write equations using average speed like I did in Lesson 10.

   I need to define my variables for the equations to make sense.

   b. Write the linear equation that represents Allen's constant speed. Make sure to include in your equation the extra time that Allen was able to walk.

   Allen's rate is $\frac{1}{60}$ miles per minute. If Allen continues walking y miles in x minutes at a constant speed, then $y = \frac{1}{60}x$. To account for the extra time that Allen gets to walk, we write the equation

   To account for the extra time, I need to add 10 minutes to Allen's time of x minutes.

   $$y = \frac{1}{60}(x + 10)$$
   $$y = \frac{1}{60}x + \frac{1}{6}$$

   When I distribute $\frac{1}{60}$ to 10, I can write it as $\frac{10}{60}$, or $\frac{1}{6}$.

   c. Write the system of linear equations that represents this situation.

   Writing a system means to write both of the equations with the bracket in front.

   $$\begin{cases} y = \frac{1}{45}x \\ y = \frac{1}{60}x + \frac{1}{6} \end{cases}$$

d. Sketch the graphs of the two equations.

I put information about Regina's walk in a table to help me graph.

Number of Minutes (x)	Miles Walked (y)
0	0
9	0.2
18	0.4

I can do the same with information about Allen's walk.

I will label the axis according to how I defined my variables. I need to label the graph of each line.

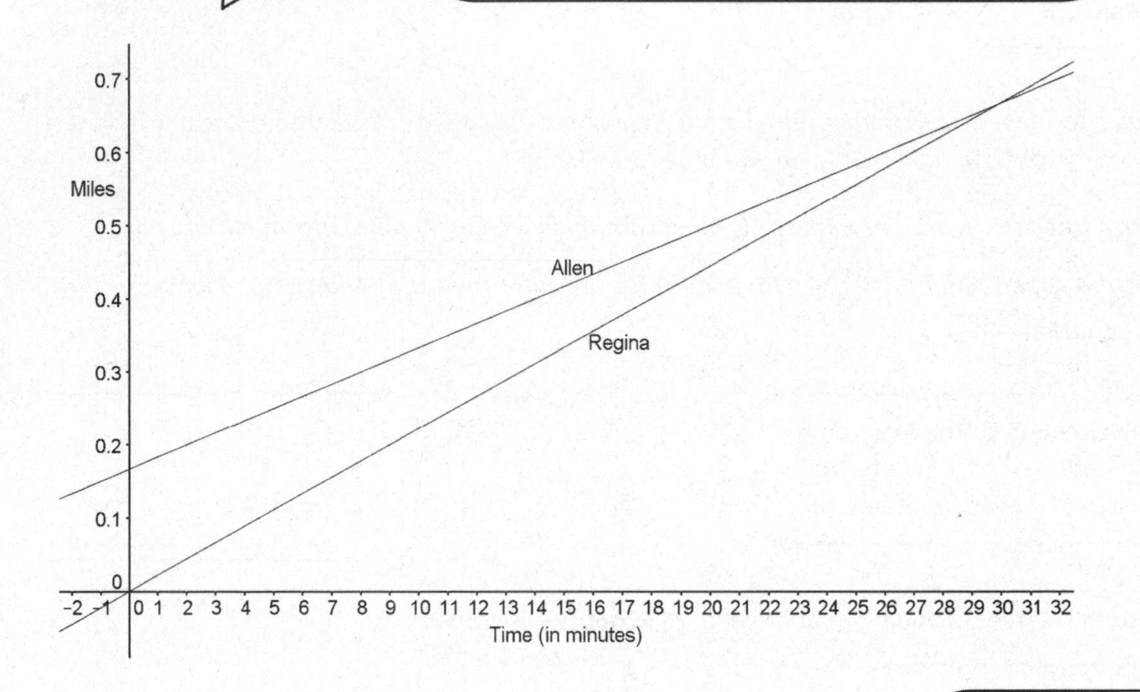

e. Will Regina ever catch up to Allen? If so, approximately when?

Yes, Regina will catch up to Allen after about 30 ***minutes or about*** 0.65 ***miles.***

I can use the graph to see at what point the graphs of the lines intersect. This will tell me when Regina will catch up to Allen.

f. At approximately what point do the graphs of the lines intersect?

The lines intersect at approximately $(30, 0.65)$.

1. Jeremy and Gerardo run at constant speeds. Jeremy can run 1 mile in 8 minutes, and Gerardo can run 3 miles in 33 minutes. Jeremy started running 10 minutes after Gerardo. Assuming they run the same path, when will Jeremy catch up to Gerardo?

   a. Write the linear equation that represents Jeremy's constant speed.

   b. Write the linear equation that represents Gerardo's constant speed. Make sure to include in your equation the extra time that Gerardo was able to run.

   c. Write the system of linear equations that represents this situation.

   d. Sketch the graphs of the two equations.

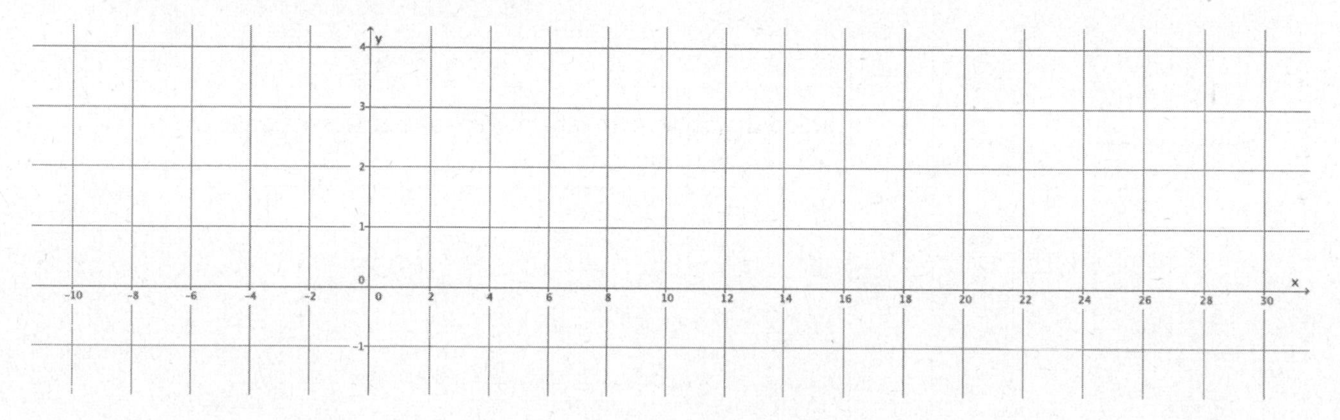

   e. Will Jeremy ever catch up to Gerardo? If so, approximately when?

   f. At approximately what point do the graphs of the lines intersect?

2. Two cars drive from town A to town B at constant speeds. The blue car travels 25 miles per hour, and the red car travels 60 miles per hour. The blue car leaves at 9:30 a.m., and the red car leaves at noon. The distance between the two towns is 150 miles.

   a. Who will get there first? Write and graph the system of linear equations that represents this situation.

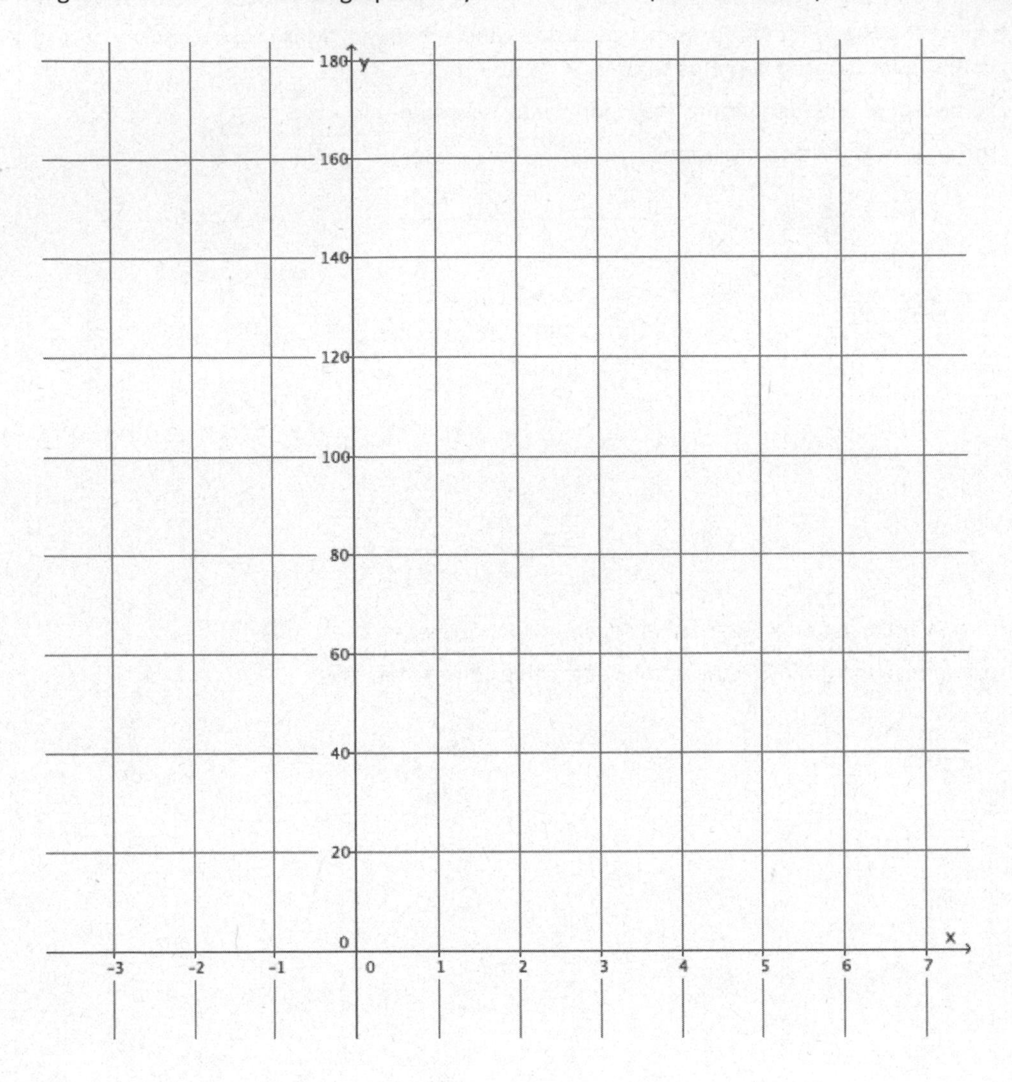

   b. At approximately what point do the graphs of the lines intersect?

Exploratory Challenge/Exercises 1–5

1. Sketch the graphs of the linear system on a coordinate plane: $\begin{cases} 2y + x = 12 \\ y = \frac{5}{6}x - 2 \end{cases}$.

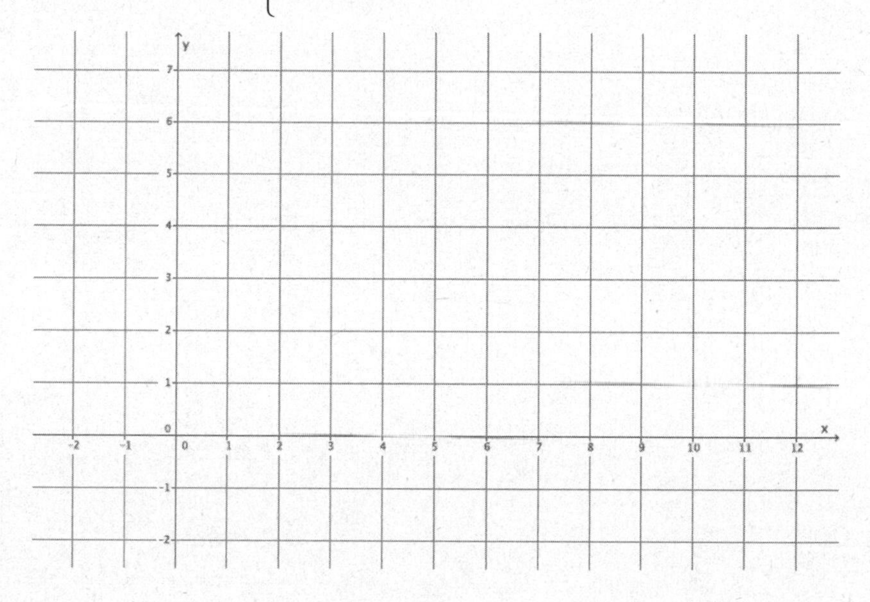

a. Name the ordered pair where the graphs of the two linear equations intersect.

b. Verify that the ordered pair named in part (a) is a solution to $2y + x = 12$.

c. Verify that the ordered pair named in part (a) is a solution to $y = \frac{5}{6}x - 2$

d. Could the point $(4, 4)$ be a solution to the system of linear equations? That is, would $(4, 4)$ make both equations true? Why or why not?

2. Sketch the graphs of the linear system on a coordinate plane: $\begin{cases} x + y = -2 \\ y = 4x + 3 \end{cases}$.

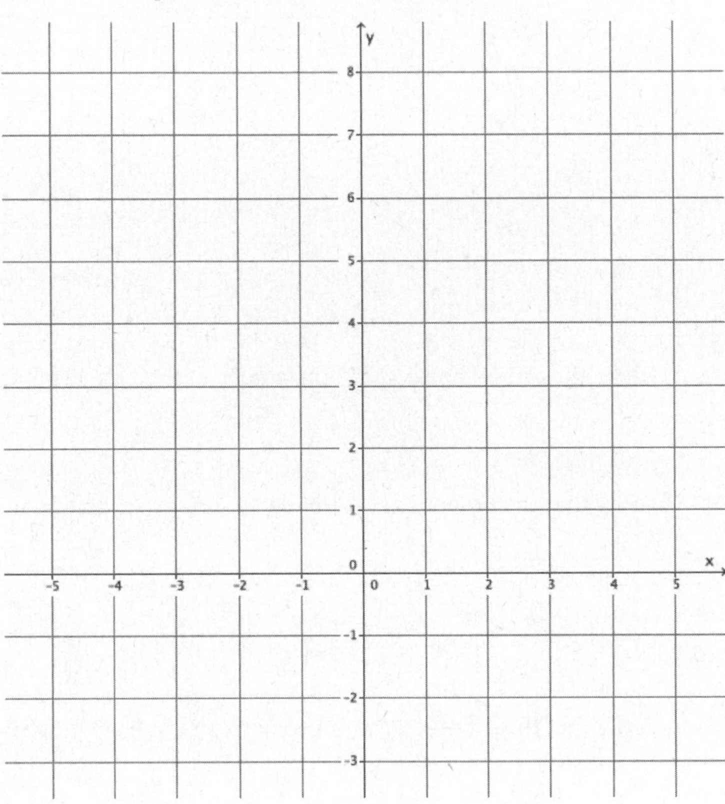

a. Name the ordered pair where the graphs of the two linear equations intersect.

b. Verify that the ordered pair named in part (a) is a solution to $x + y = -2$.

c. Verify that the ordered pair named in part (a) is a solution to $y = 4x + 3$.

d. Could the point $(-4, 2)$ be a solution to the system of linear equations? That is, would $(-4, 2)$ make both equations true? Why or why not?

3. Sketch the graphs of the linear system on a coordinate plane: $\begin{cases} 3x + y = -3 \\ -2x + y = 2 \end{cases}$.

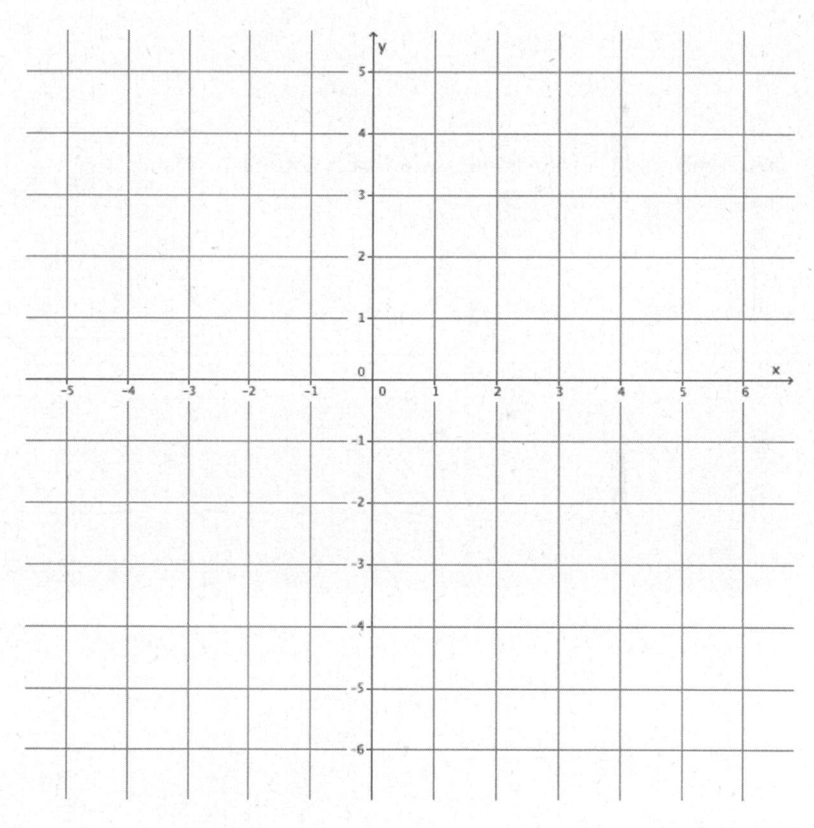

a. Name the ordered pair where the graphs of the two linear equations intersect.

b. Verify that the ordered pair named in part (a) is a solution to $3x + y = -3$.

c. Verify that the ordered pair named in part (a) is a solution to $-2x + y = 2$.

d. Could the point $(1, 4)$ be a solution to the system of linear equations? That is, would $(1, 4)$ make both equations true? Why or why not?

4. Sketch the graphs of the linear system on a coordinate plane: $\begin{cases} 2x - 3y = 18 \\ 2x + y = 2 \end{cases}$.

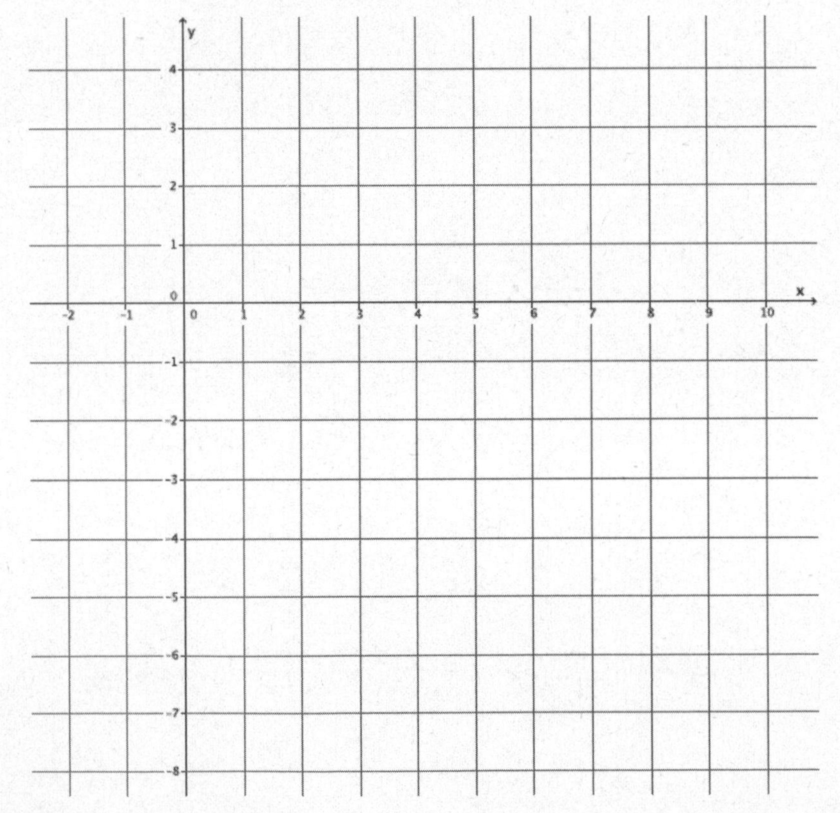

a. Name the ordered pair where the graphs of the two linear equations intersect.

b. Verify that the ordered pair named in part (a) is a solution to $2x - 3y = 18$.

c. Verify that the ordered pair named in part (a) is a solution to $2x + y = 2$.

d. Could the point $(3, -1)$ be a solution to the system of linear equations? That is, would $(3, -1)$ make both equations true? Why or why not?

5. Sketch the graphs of the linear system on a coordinate plane: $\begin{cases} y - x = 3 \\ y = -4x - 2 \end{cases}$.

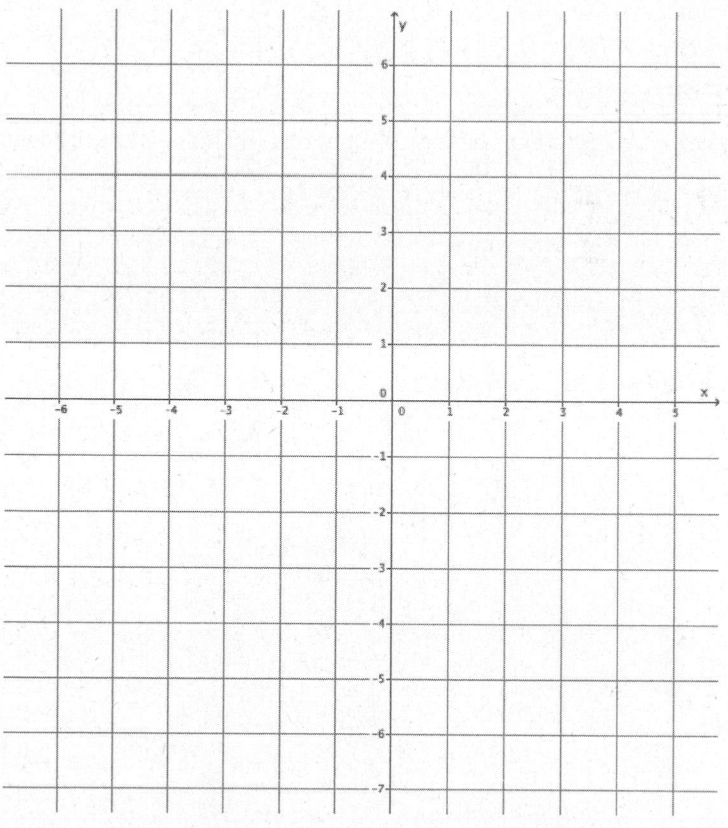

a. Name the ordered pair where the graphs of the two linear equations intersect.

b. Verify that the ordered pair named in part (a) is a solution to $y - x = 3$.

c. Verify that the ordered pair named in part (a) is a solution to $y = -4x - 2$.

d. Could the point $(-2, 6)$ be a solution to the system of linear equations? That is, would $(-2, 6)$ make both equations true? Why or why not?

Exercise 6

6. Write two different systems of equations with $(1, -2)$ as the solution.

Lesson Summary

When the graphs of a system of linear equations are sketched, and if they are not parallel lines, then the point of intersection of the lines of the graph represents the solution to the system. Two distinct lines intersect at most at one point, if they intersect. The coordinates of that point (x, y) represent values that make both equations of the system true.

Example: The system $\begin{cases} x + y = 3 \\ x - y = 5 \end{cases}$ graphs as shown below.

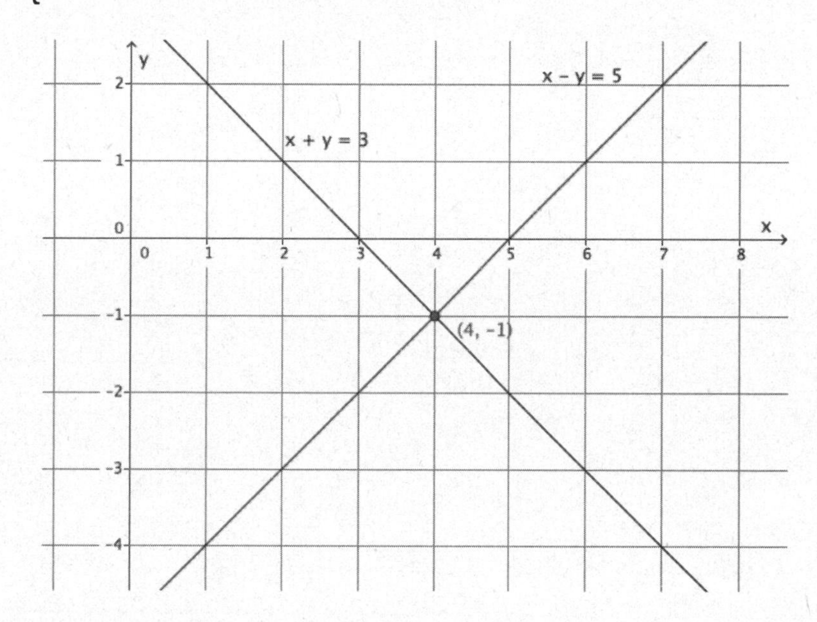

The lines intersect at $(4, -1)$. That means the equations in the system are true when $x = 4$ and $y = -1$.

$$x + y = 3$$
$$4 + (-1) = 3$$
$$3 = 3$$

$$x - y = 5$$
$$4 - (-1) = 5$$
$$5 = 5$$

Name ____ Date ____________________

Sketch the graphs of the linear system on a coordinate plane: $\begin{cases} 2x - y = -1 \\ y = 5x - 5 \end{cases}$.

a. Name the ordered pair where the graphs of the two linear equations intersect.

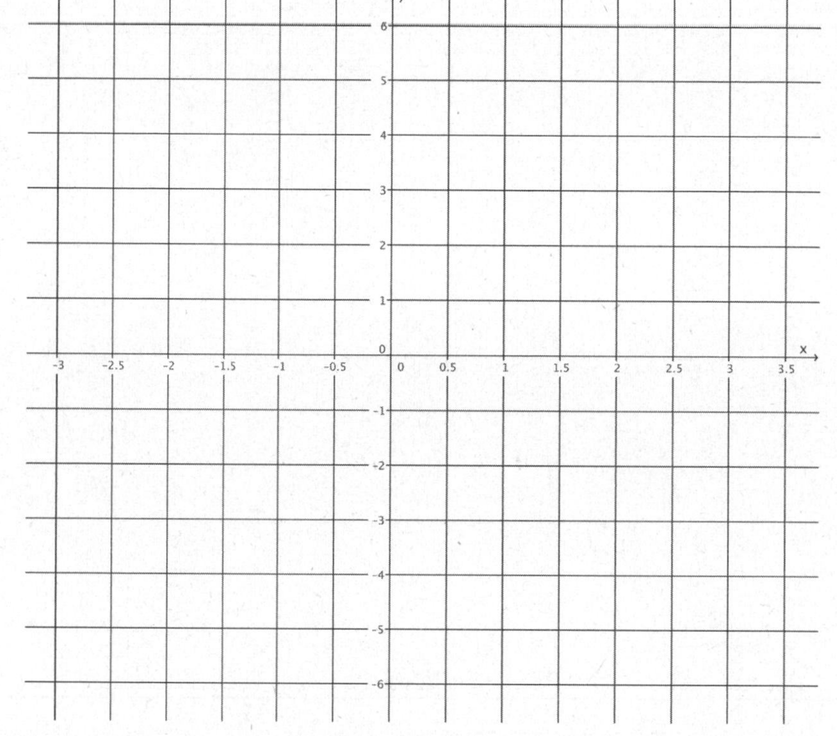

b. Verify that the ordered pair named in part (a) is a solution to $2x - y = -1$.

c. Verify that the ordered pair named in part (a) is a solution to $y = 5x - 5$.

1. Sketch the graphs of the linear system on a coordinate plane: $\begin{cases} y = -\frac{1}{9}x - 3 \\ -2x + 3y = 12 \end{cases}$

> I will use the slope and y-intercept to help me graph the equation of this line.

For the equation $: y = -\frac{1}{9}x - 3$

The slope is $-\frac{1}{9}$, ***and the y-intercept is*** $(0, -3)$.

> Since this equation is in standard form, I will fix $x = 0$ to find y (the y-intercept) and fix $y = 0$ to find x (the x-intercept) to help me graph the line of this equation.

For the equation $-2x + 3y = 12$;

$$-2(0) + 3y = 12$$
$$3y = 12$$
$$y = 4$$

The y-intercept is $(0, 4)$.

$$-2x + 3(0) = 12$$
$$-2x = 12$$
$$x = -6$$

The x-intercept is $(-6, 0)$.

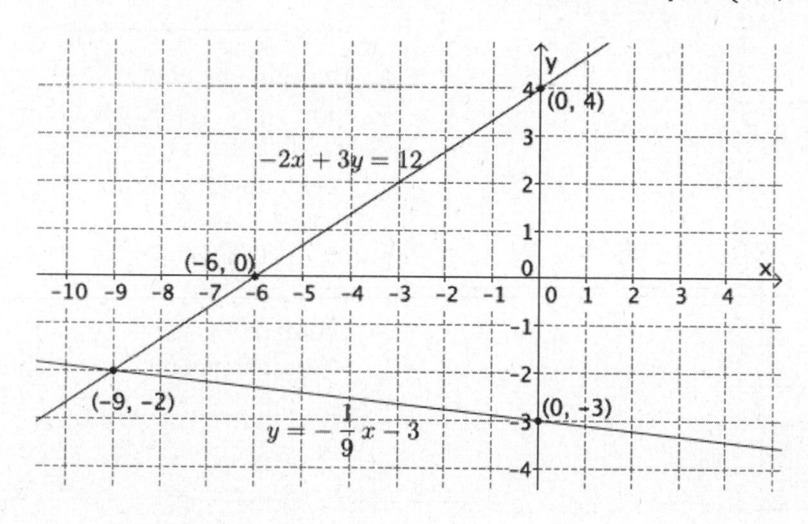

> To locate where the graphs of the lines intersect, I will use graph paper so that I can be as accurate as possible.

a. Name the ordered pair where the graphs of the two linear equations intersect.

$(-9, -2)$

b. Verify that the ordered pair named in part (a) is a solution to $y = -\frac{1}{9}x - 3$.

$$-2 = -\frac{1}{9}(-9) - 3$$
$$-2 = 1 - 3$$
$$-2 = -2$$

The left and right sides of the equation are equal.

c. Verify that the ordered pair named in part (a) is a solution to $-2x + 3y = 12$.

$$-2(-9) + 3(-2) = 12$$
$$18 - 6 = 12$$
$$12 = 12$$

The left and right sides of the equation are equal.

If the solutions I found on the graph of the lines satisfies both equations, then I have found an ordered pair of the system.

2. Sketch the graphs of the linear system on a coordinate plane: $\begin{cases} y = 5x + 7 \\ y = -3 \end{cases}$

The equation of the form $y = c$, where c is a constant, is a horizontal line passing though the point $(0, c)$. I know that -3 will be the y-coordinate of my intersection point.

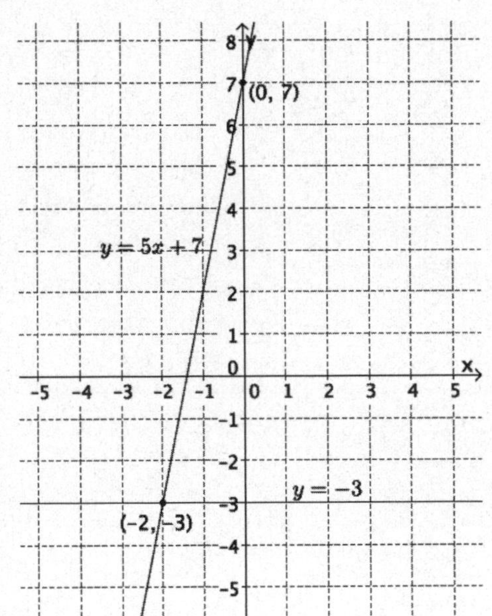

I can write the slope of 5 with the denominator of 1 as the fraction $\frac{5}{1}$.

For the equation $y = 5x + 7$:

The slope is $\frac{5}{1}$, and the y-intercept is $(0, 7)$.

I need to verify that the ordered pair is a solution to both of the equations.

Name the ordered pair where the graphs of the two linear equations intersect.

$(-2, -3)$

1. Sketch the graphs of the linear system on a coordinate plane: $\begin{cases} y = \frac{1}{3}x + 1 \\ y = -3x + 11 \end{cases}$.
   a. Name the ordered pair where the graphs of the two linear equations intersect.
   b. Verify that the ordered pair named in part (a) is a solution to $y = \frac{1}{3}x + 1$.
   c. Verify that the ordered pair named in part (a) is a solution to $y = -3x + 11$.

2. Sketch the graphs of the linear system on a coordinate plane: $\begin{cases} y = \frac{1}{2}x + 4 \\ x + 4y = 4 \end{cases}$.
   a. Name the ordered pair where the graphs of the two linear equations intersect.
   b. Verify that the ordered pair named in part (a) is a solution to $y = \frac{1}{2}x + 4$.
   c. Verify that the ordered pair named in part (a) is a solution to $x + 4y = 4$.

3. Sketch the graphs of the linear system on a coordinate plane: $\begin{cases} y = 2 \\ x + 2y = 10 \end{cases}$.
   a. Name the ordered pair where the graphs of the two linear equations intersect.
   b. Verify that the ordered pair named in part (a) is a solution to $y = 2$.
   c. Verify that the ordered pair named in part (a) is a solution to $x + 2y = 10$.

4. Sketch the graphs of the linear system on a coordinate plane: $\begin{cases} -2x + 3y = 18 \\ 2x + 3y = 6 \end{cases}$.
   a. Name the ordered pair where the graphs of the two linear equations intersect.
   b. Verify that the ordered pair named in part (a) is a solution to $-2x + 3y = 18$.
   c. Verify that the ordered pair named in part (a) is a solution to $2x + 3y = 6$.

5. Sketch the graphs of the linear system on a coordinate plane: $\begin{cases} x + 2y = 2 \\ y = \frac{2}{3}x - 6 \end{cases}$.
   a. Name the ordered pair where the graphs of the two linear equations intersect.
   b. Verify that the ordered pair named in part (a) is a solution to $x + 2y = 2$.
   c. Verify that the ordered pair named in part (a) is a solution to $y = \frac{2}{3}x - 6$.

6. Without sketching the graph, name the ordered pair where the graphs of the two linear equations intersect.

$$\begin{cases} x = 2 \\ y = -3 \end{cases}$$

Exercises

1. Sketch the graphs of the system. $\begin{cases} y = \frac{2}{3}x + 4 \\ y = \frac{4}{6}x - 3 \end{cases}$

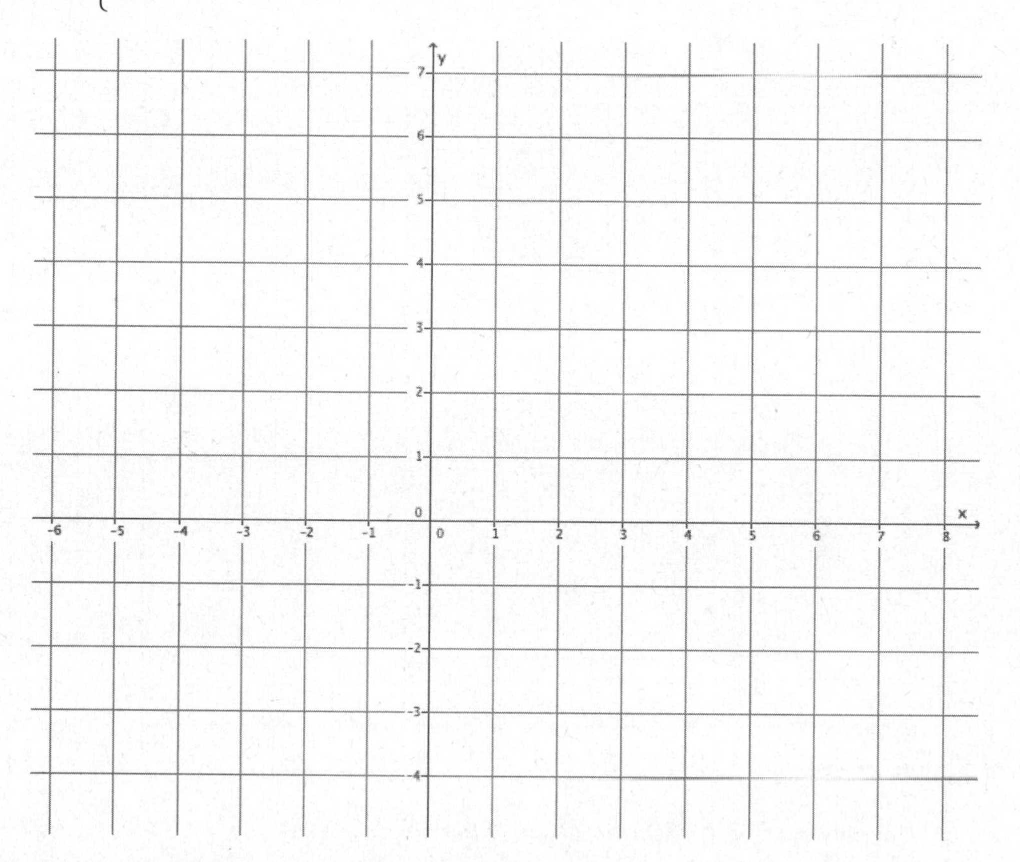

a. Identify the slope of each equation. What do you notice?

b. Identify the y-intercept point of each equation. Are the y-intercept points the same or different?

2. Sketch the graphs of the system. $\begin{cases} y = -\frac{5}{4}x + 7 \\ y = -\frac{5}{4}x + 2 \end{cases}$

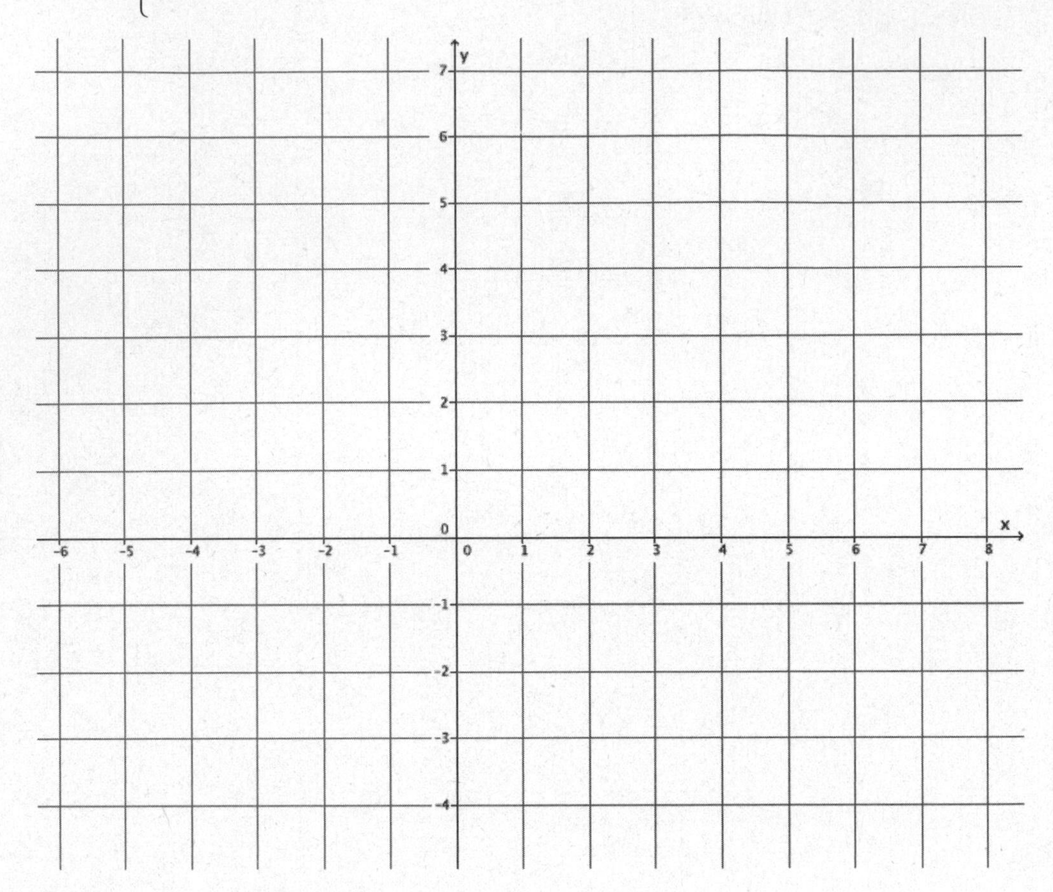

a. Identify the slope of each equation. What do you notice?

b. Identify the y-intercept point of each equation. Are the y-intercept points the same or different?

3. Sketch the graphs of the system. $\begin{cases} y = 2x - 5 \\ y = 2x - 1 \end{cases}$

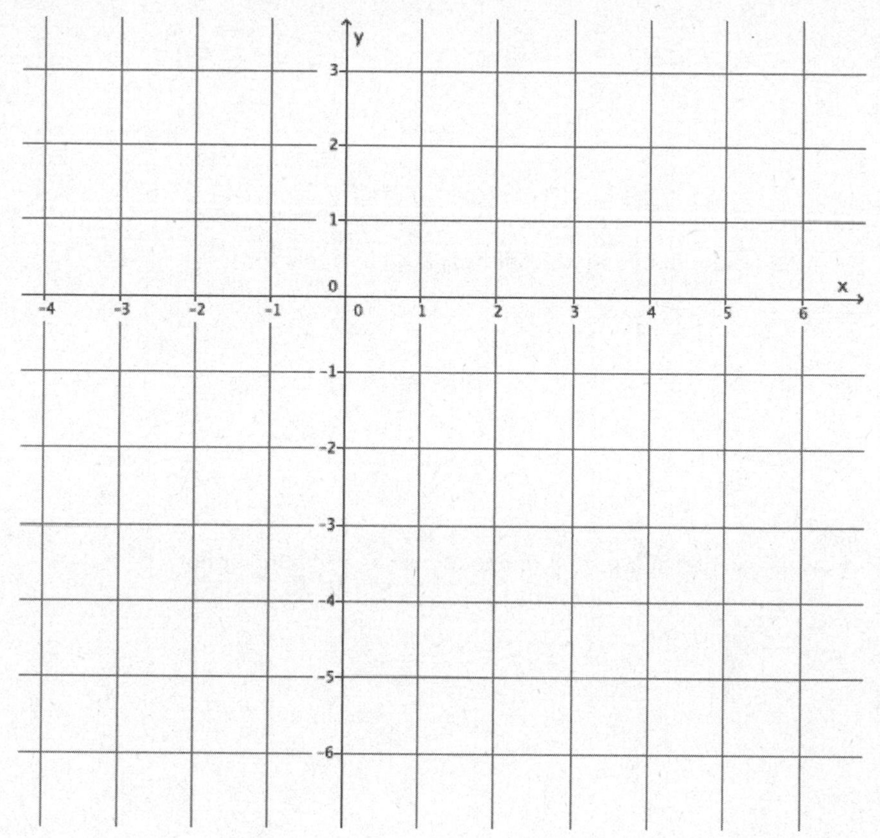

a. Identify the slope of each equation. What do you notice?

b. Identify the y-intercept point of each equation. Are the y-intercept points the same or different?

4. Write a system of equations that has no solution.

5. Write a system of equations that has $(2, 1)$ as a solution.

6. How can you tell if a system of equations has a solution or not?

7. Does the system of linear equations shown below have a solution? Explain.

$$\begin{cases} 6x - 2y = 5 \\ 4x - 3y = 5 \end{cases}$$

8. Does the system of linear equations shown below have a solution? Explain.

$$\begin{cases} -2x + 8y = 14 \\ x = 4y + 1 \end{cases}$$

9. Does the system of linear equations shown below have a solution? Explain.

$$\begin{cases} 12x + 3y = -2 \\ 4x + y = 7 \end{cases}$$

10. Genny babysits for two different families. One family pays her \$6 each hour and a bonus of \$20 at the end of the night. The other family pays her \$3 every half hour and a bonus of \$25 at the end of the night. Write and solve the system of equations that represents this situation. At what number of hours do the two families pay the same for babysitting services from Genny?

Lesson Summary

By definition, parallel lines do not intersect; therefore, a system of linear equations whose graphs are parallel lines will have no solution.

Parallel lines have the same slope but no common point. One can verify that two lines are parallel by comparing their slopes and their y-intercept points.

Name ____________________ Date ____________

Does each system of linear equations have a solution? Explain your answer.

1. $\begin{cases} y = \frac{5}{4}x - 3 \\ y + 2 = \frac{5}{4}x \end{cases}$

2. $\begin{cases} y = \frac{2}{3}x - 5 \\ 4x - 8y = 11 \end{cases}$

3. $\begin{cases} \frac{1}{3}x + y = 8 \\ x + 3y = 12 \end{cases}$

I need to determine if the graphs of the lines are parallel. Parallel lines do not intersect, which means parallel lines have no solution.

Answer Problems 1–2 without graphing the equations.

1. Does the system of linear equations shown below have a solution? Explain.

$$\begin{cases} 2x + 5y = 9 \\ -4x - 10y = 4 \end{cases}$$

Standard form is $ax + by = c$, where a, b, c are constants and a and b are not both zero.

I learned from Lesson 23 that when equations are written in standard form, I know the slope is $m = -\frac{a}{b}$, and the y-intercept is $\frac{c}{b}$.

No, this system does not have a solution. The slope of the first equation is $-\frac{2}{5}$, and the slope of the second equation is $-\frac{4}{10}$, which is equivalent to $-\frac{2}{5}$. Since the slopes are the same and the lines are distinct, these equations will graph as parallel lines. Parallel lines never intersect, which means this system has no solution.

2. Does the system of linear equations shown below have a solution? Explain.

$$\begin{cases} \frac{7}{4}x + 2 = y \\ x + 2y = 4 \end{cases}$$

If the slopes are different, these equations will graph as non-parallel lines and intersect at some point. That means they will have a solution.

The first equation is written in slope-intercept form. The slope is $\frac{7}{4}$.

Yes, this system does have a solution. The slope of the first equation is $\frac{7}{4}$, and the slope of the second equation is $-\frac{1}{2}$. Since the slopes are different, these equations will graph as non-parallel lines, which means they will intersect at some point

3. Given the graphs of a system of linear equations below, is there a solution to the system that we cannot see on this portion of the coordinate plane? That is, will the lines intersect somewhere on the plane not represented in the picture? Explain.

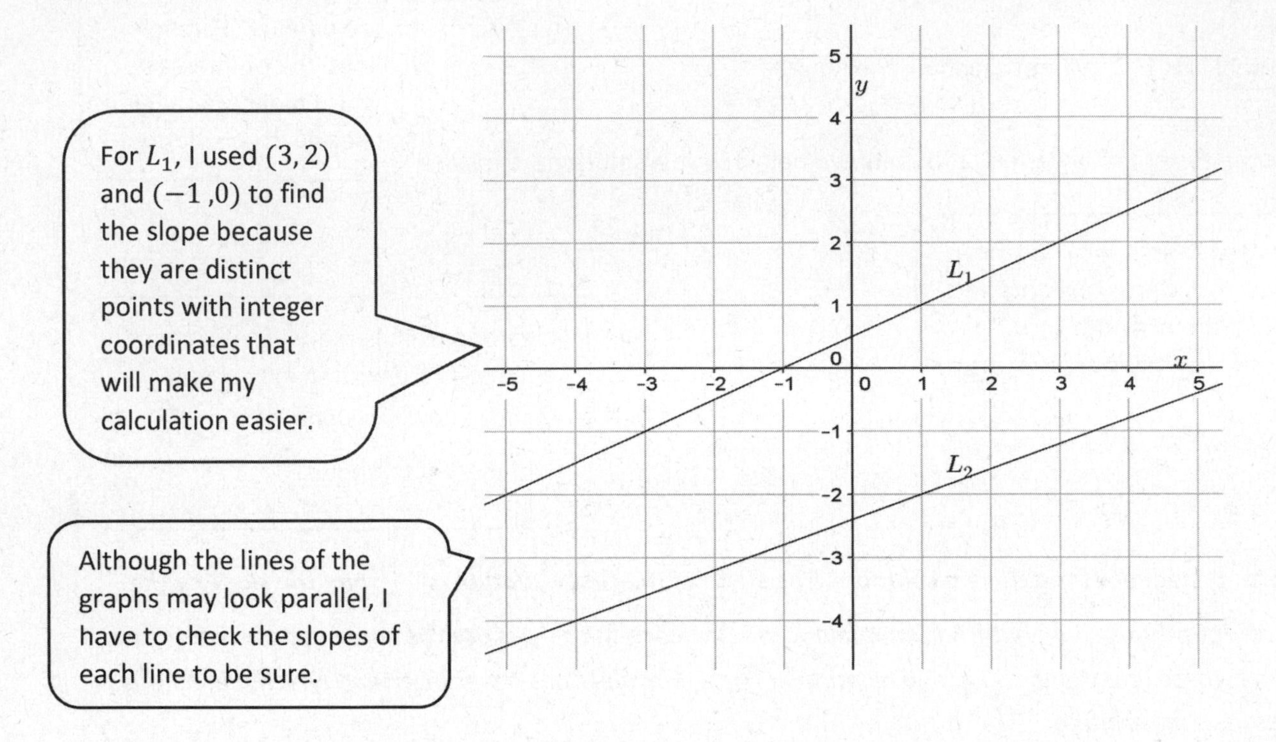

The slope of L_1 is $\frac{1}{2}$, and the slope of L_2 is $\frac{2}{5}$. Since the slopes are different, these lines are nonparallel lines, which means they will intersect at some point. Therefore, the system of linear equations whose graphs are the given lines will have a solution.

Answer Problems 1–5 without graphing the equations.

1. Does the system of linear equations shown below have a solution? Explain.

$$\begin{cases} 2x + 5y = 9 \\ -4x - 10y = 4 \end{cases}$$

2. Does the system of linear equations shown below have a solution? Explain.

$$\begin{cases} \frac{3}{4}x - 3 = y \\ 4x - 3y = 5 \end{cases}$$

3. Does the system of linear equations shown below have a solution? Explain.

$$\begin{cases} x + 7y = 8 \\ 7x - y = -2 \end{cases}$$

4. Does the system of linear equations shown below have a solution? Explain.

$$\begin{cases} y = 5x + 12 \\ 10x - 2y = 1 \end{cases}$$

5. Does the system of linear equations shown below have a solution? Explain.

$$\begin{cases} y = \frac{5}{3}x + 15 \\ 5x - 3y = 6 \end{cases}$$

6. Given the graphs of a system of linear equations below, is there a solution to the system that we cannot see on this portion of the coordinate plane? That is, will the lines intersect somewhere on the plane not represented in the picture? Explain.

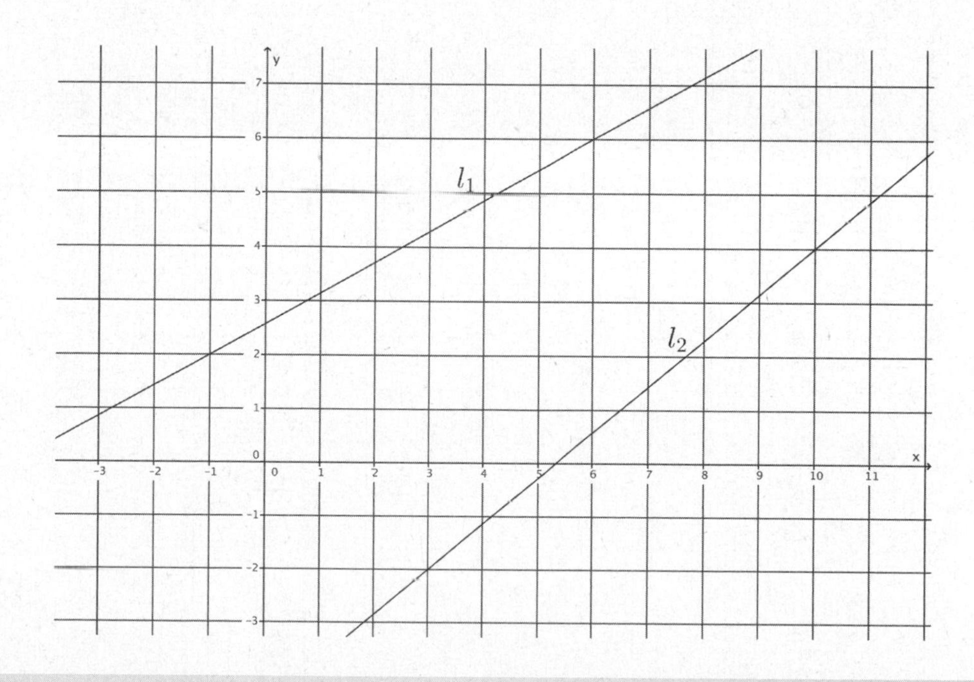

7. Given the graphs of a system of linear equations below, is there a solution to the system that we cannot see on this portion of the coordinate plane? That is, will the lines intersect somewhere on the plane not represented in the picture? Explain.

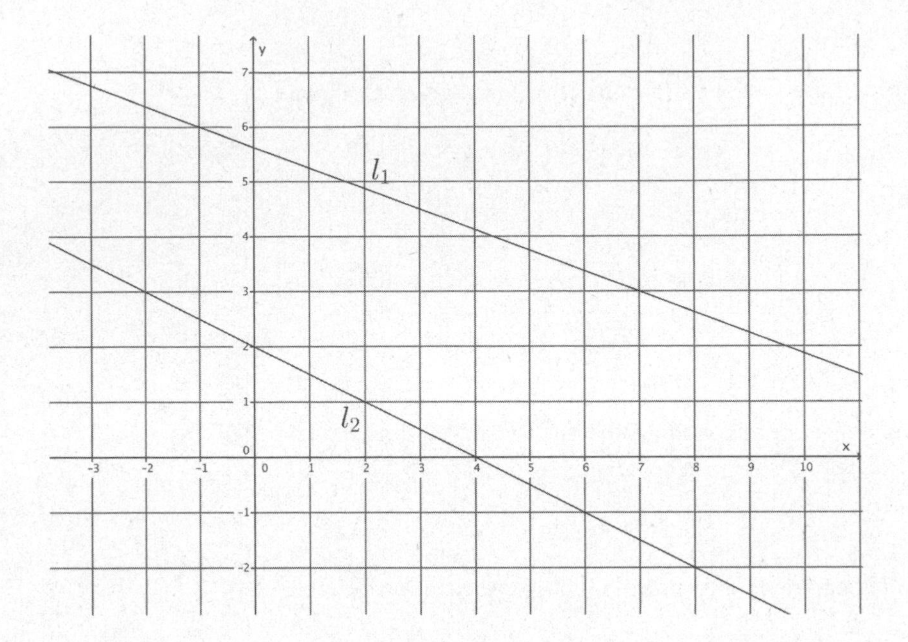

8. Given the graphs of a system of linear equations below, is there a solution to the system that we cannot see on this portion of the coordinate plane? That is, will the lines intersect somewhere on the plane not represented in the picture? Explain.

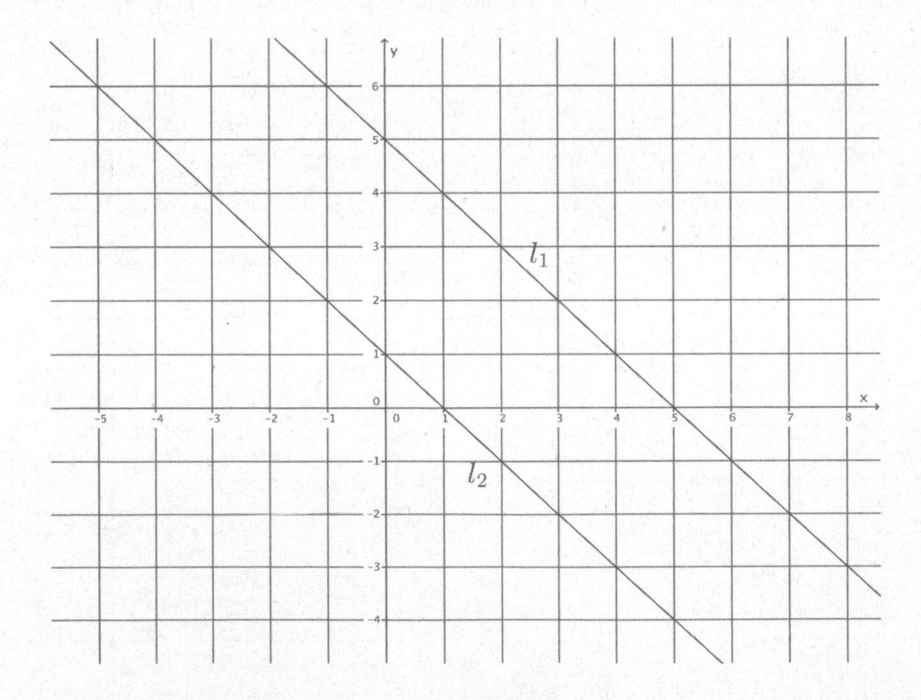

9. Given the graphs of a system of linear equations below, is there a solution to the system that we cannot see on this portion of the coordinate plane? That is, will the lines intersect somewhere on the plane not represented in the picture? Explain.

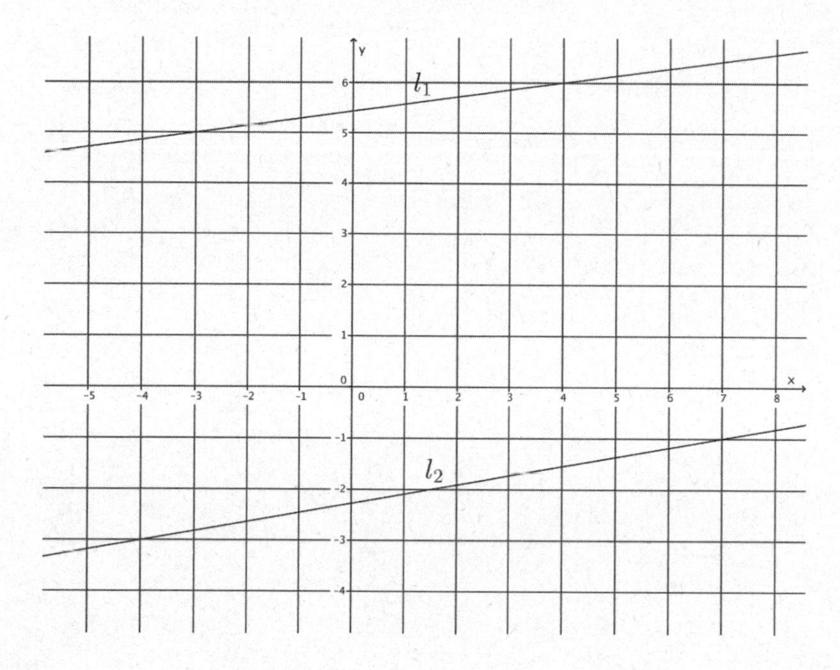

10. Given the graphs of a system of linear equations below, is there a solution to the system that we cannot see on this portion of the coordinate plane? That is, will the lines intersect somewhere on the plane not represented in the picture? Explain.

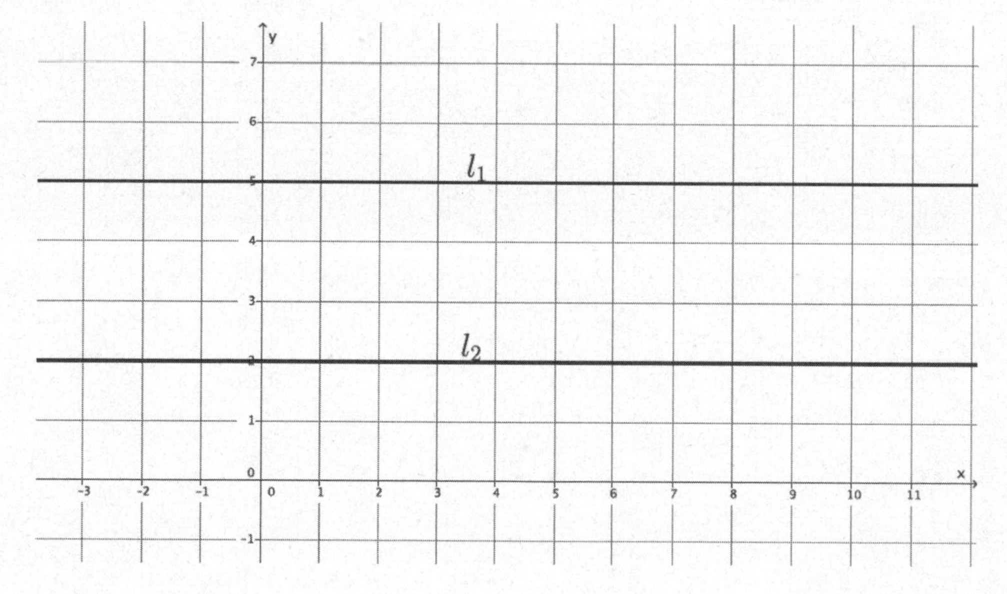

Exercises

Determine the nature of the solution to each system of linear equations.

1. $\begin{cases} 3x + 4y = 5 \\ y = -\frac{3}{4}x + 1 \end{cases}$

2. $\begin{cases} 7x + 2y = -4 \\ x - y = 5 \end{cases}$

3. $\begin{cases} 9x + 6y = 3 \\ 3x + 2y = 1 \end{cases}$

Determine the nature of the solution to each system of linear equations. If the system has a solution, find it algebraically, and then verify that your solution is correct by graphing.

4. $\begin{cases} 3x + 3y = -21 \\ x + y = -7 \end{cases}$

5. $\begin{cases} y = \frac{3}{2}x - 1 \\ 3y = x + 2 \end{cases}$

6. $\begin{cases} x = 12y - 4 \\ x = 9y + 7 \end{cases}$

7. Write a system of equations with $(4, -5)$ as its solution.

Lesson Summary

A system of linear equations can have a unique solution, no solution, or infinitely many solutions.

Systems with a unique solution are comprised of two linear equations whose graphs have different slopes; that is, their graphs in a coordinate plane will be two distinct lines that intersect at only one point.

Systems with no solutions are comprised of two linear equations whose graphs have the same slope but different y-intercept points; that is, their graphs in a coordinate plane will be two parallel lines (with no intersection).

Systems with infinitely many solutions are comprised of two linear equations whose graphs have the same slope and the same y-intercept point; that is, their graphs in a coordinate plane are the same line (i.e., every solution to one equation will be a solution to the other equation).

A system of linear equations can be solved using a substitution method. That is, if two expressions are equal to the same value, then they can be written equal to one another.

Example:

$$\begin{cases} y = 5x - 8 \\ y = 6x + 3 \end{cases}$$

Since both equations in the system are equal to y, we can write the equation $5x - 8 = 6x + 3$ and use it to solve for x and then the system.

Example:

$$\begin{cases} 3x = 4y + 2 \\ x = y + 5 \end{cases}$$

Multiply each term of the equation $x = y + 5$ by 3 to produce the equivalent equation $3x = 3y + 15$. As in the previous example, since both equations equal $3x$, we can write $4y + 2 = 3y + 15$. This equation can be used to solve for y and then the system.

Name ____________________ Date ____________

Determine the nature of the solution to each system of linear equations. If the system has a solution, then find it without graphing.

1. $\begin{cases} y = \frac{1}{2}x + \frac{5}{2} \\ x - 2y = 7 \end{cases}$

2. $\begin{cases} y = \frac{2}{3}x + 4 \\ 2y + \frac{1}{2}x = 2 \end{cases}$

3. $\begin{cases} y = 3x - 2 \\ -3x + y = -2 \end{cases}$

Determine the nature of the solution to each system of linear equations. If the system has a solution, find it algebraically, and then verify that your solution is correct by graphing.

1. $\begin{cases} y = -\frac{4}{5}x + 9 \\ 4x + 5y = 9 \end{cases}$

If the equations have the same slope and different y-intercepts, then the equations graph as parallel lines, which means the system doesn't have a solution.

The slopes of these two equations are the same, and the y-intercepts are different, which means they graph as parallel lines. Therefore, this system will have no solutions.

2. $\begin{cases} 2x - 3y = 12 \\ y = \frac{2}{3}x - 4 \end{cases}$

I notice that if I multiply the second equation by 3, the result is $3y = 2x - 12$. When I use my properties of equality, I see the second equation is the same as the first. This means that I have the same line; therefore, I have infinitely many solutions.

These equations define the same line. Therefore, this system will have infinitely many solutions.

3. $\begin{cases} y = 5x - 1 \\ y = 11x + 2 \end{cases}$

Since the equations are both equal to y, I can use substitution, and write the equations equal to each other and solve for x.

$$5x - 1 = 11x + 2$$
$$-3 = 6x$$
$$-\frac{1}{2} = x$$

Once I solve for x, then I can use substitution again in either equation and solve for y.

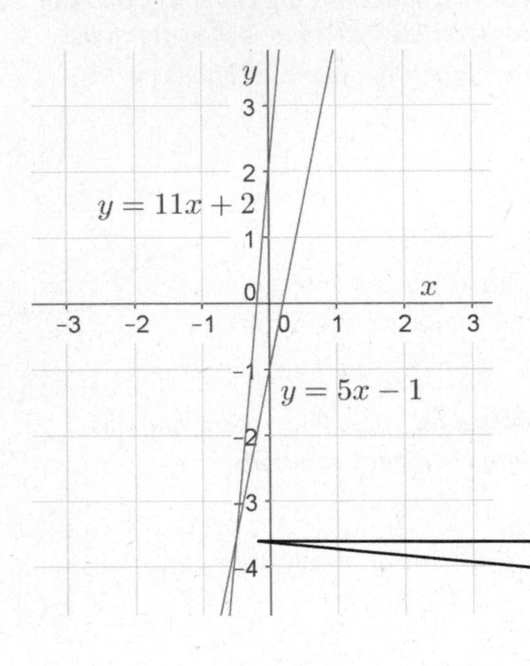

$$y = 5\left(-\frac{1}{2}\right) - 1$$
$$y = -\frac{5}{2} - 1$$
$$y = -\frac{7}{2}$$

The solution is $\left(-\frac{1}{2}, -\frac{7}{2}\right)$

I see that the graphs of the lines intersect at $\left(-\frac{1}{2}, -\frac{7}{2}\right)$.

4. $\begin{cases} 6x - 2 = y \\ 2y = 2x + 5 \end{cases}$

I can multiply the first equation by 2 to produce an equivalent equation, namely $12x - 4 = 2y$. Now that both equations are equal to $2y$, the expressions $12x - 4$ and $2x + 5$ can be written equal to one another.

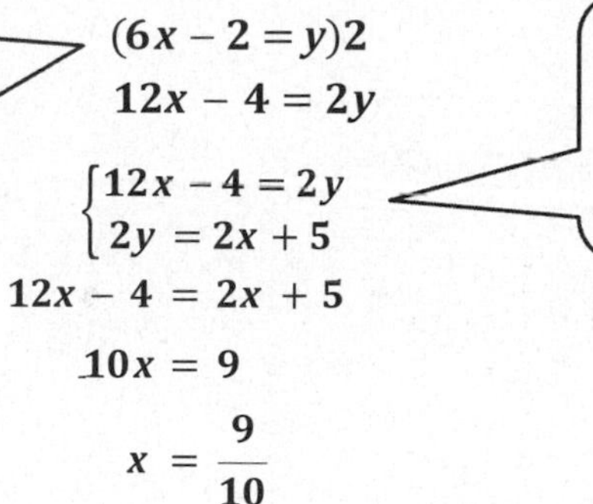

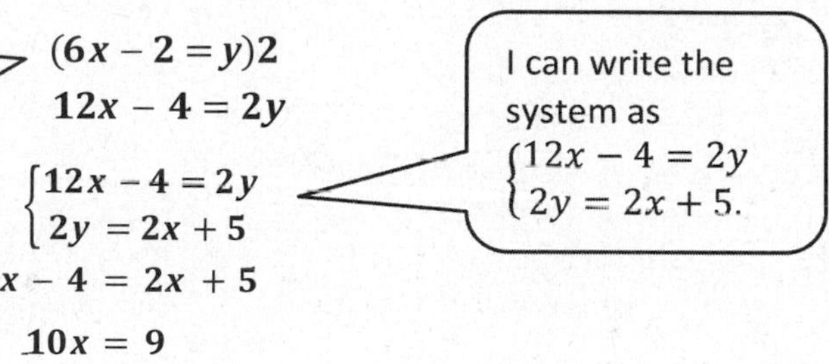

$$(6x - 2 = y)2$$
$$12x - 4 = 2y$$

I can write the system as $\begin{cases} 12x - 4 = 2y \\ 2y = 2x + 5. \end{cases}$

$$\begin{cases} 12x - 4 = 2y \\ 2y = 2x + 5 \end{cases}$$
$$12x - 4 = 2x + 5$$
$$10x = 9$$
$$x = \frac{9}{10}$$

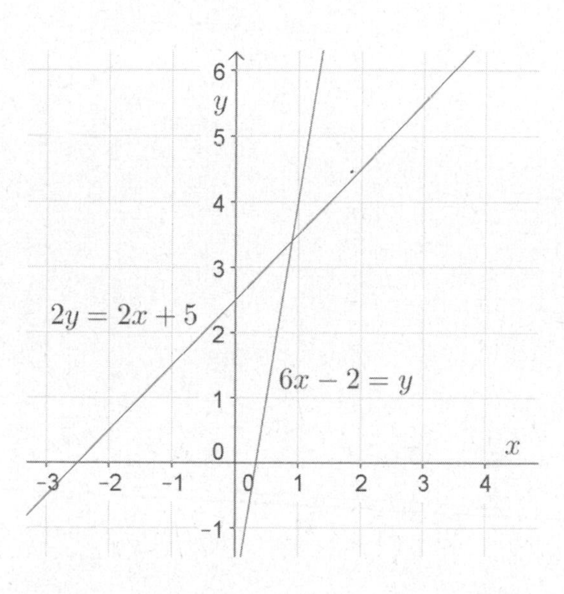

$$6\left(\frac{9}{10}\right) - 2 = y$$
$$\frac{54}{10} - 2 = y$$
$$\frac{17}{5} = y$$

The solution is $\left(\frac{9}{10}, \frac{17}{5}\right)$.

Determine the nature of the solution to each system of linear equations. If the system has a solution, find it algebraically, and then verify that your solution is correct by graphing.

1. $\begin{cases} y = \frac{3}{7}x - 8 \\ 3x - 7y = 1 \end{cases}$

2. $\begin{cases} 2x - 5 = y \\ -3x - 1 = 2y \end{cases}$

3. $\begin{cases} x = 6y + 7 \\ x = 10y + 2 \end{cases}$

4. $\begin{cases} 5y = \frac{15}{4}x + 25 \\ y = \frac{3}{4}x + 5 \end{cases}$

5. $\begin{cases} x + 9 = y \\ x = 4y - 6 \end{cases}$

6. $\begin{cases} 3y = 5x - 15 \\ 3y = 13x - 2 \end{cases}$

7. $\begin{cases} 6x - 7y = \frac{1}{2} \\ 12x - 14y = 1 \end{cases}$

8. $\begin{cases} 5x - 2y = 6 \\ -10x + 4y = -14 \end{cases}$

9. $\begin{cases} y = \frac{3}{2}x - 6 \\ 2y = 7 - 4x \end{cases}$

10. $\begin{cases} 7x - 10 = y \\ y = 5x + 12 \end{cases}$

11. Write a system of linear equations with $(-3, 9)$ as its solution.

Example 1

Use what you noticed about adding equivalent expressions to solve the following system by elimination:

$$\begin{cases} 6x - 5y = 21 \\ 2x + 5y = -5 \end{cases}$$

Example 2

Solve the following system by elimination:

$$\begin{cases} -2x + 7y = 5 \\ 4x - 2y = 14 \end{cases}$$

Example 3

Solve the following system by elimination:

$$\begin{cases} 7x - 5y = -2 \\ 3x - 3y = 7 \end{cases}$$

Exercises

Each of the following systems has a solution. Determine the solution to the system by eliminating one of the variables. Verify the solution using the graph of the system.

1. $\begin{cases} 6x - 7y = -10 \\ 3x + 7y = -8 \end{cases}$

2. $\begin{cases} x - 4y = 7 \\ 5x + 9y = 6 \end{cases}$

3. $\begin{cases} 2x - 3y = -5 \\ 3x + 5y = 1 \end{cases}$

Lesson Summary

Systems of linear equations can be solved by eliminating one of the variables from the system. One way to eliminate a variable is by setting both equations equal to the same variable and then writing the expressions equal to one another.

Example: Solve the system $\begin{cases} y = 3x - 4 \\ y = 2x + 1 \end{cases}$.

Since the expressions $3x - 4$ and $2x + 1$ are both equal to y, they can be set equal to each other and the new equation can be solved for x:

$$3x - 4 = 2x + 1$$

Another way to eliminate a variable is by multiplying each term of an equation by the same constant to make an equivalent equation. Then, use the equivalent equation to eliminate one of the variables and solve the system.

Example: Solve the system $\begin{cases} 2x + y = 8 \\ x + y = 10 \end{cases}$.

Multiply the second equation by -2 to eliminate the x.

$$-2(x + y = 10)$$
$$-2x - 2y = -20$$

Now we have the system $\begin{cases} -2x + y = 8 \\ -2x - 2y = -20 \end{cases}$.

When the equations are added together, the x is eliminated.

$$2x + y - 2x - 2y = 8 + (-20)$$
$$y - 2y = 8 + (-20)$$

Once a solution has been found, verify the solution graphically or by substitution.

Name ______________________________ Date ______________

Determine the solution, if it exists, for each system of linear equations. Verify your solution on the coordinate plane.

1. $\begin{cases} y = 3x - 5 \\ y = -3x + 7 \end{cases}$

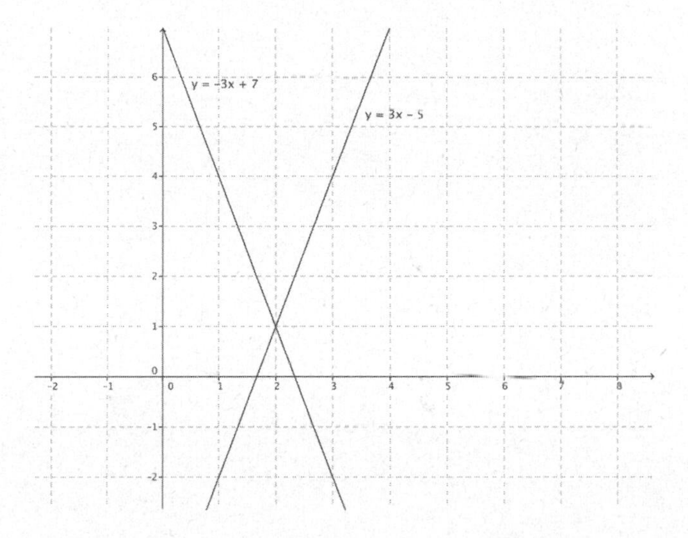

2. $\begin{cases} y = -4x + 6 \\ 2x - y = 11 \end{cases}$

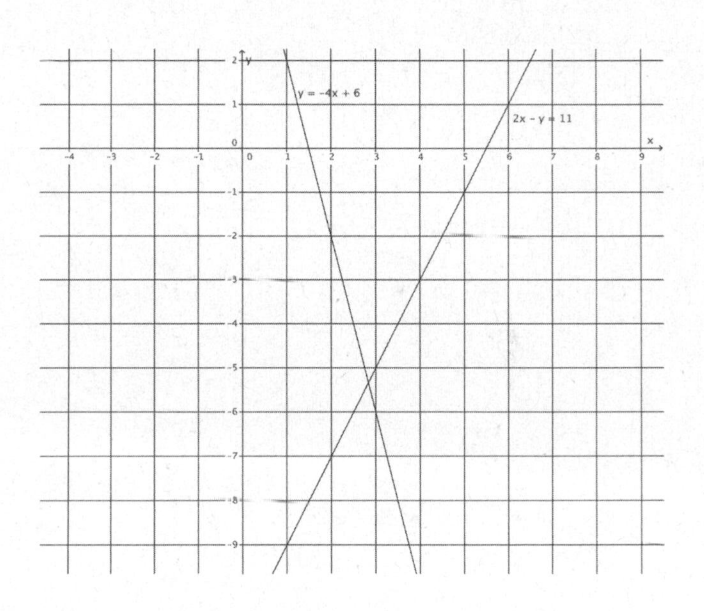

I can determine if the solutions exist by checking the slope and the y-intercepts like I did in the previous lesson's Problem Set.

Determine the solution, if it exists, for each system of linear equations. Verify your solution on the coordinate plane.

1. $\begin{cases} 5x + 3y = -2 \\ 2x - y = 6 \end{cases}$

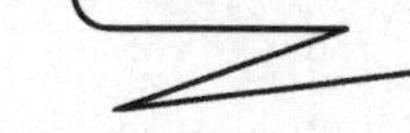

If I multiply the second equation by 3, then I will eliminate x and be able to solve for y.

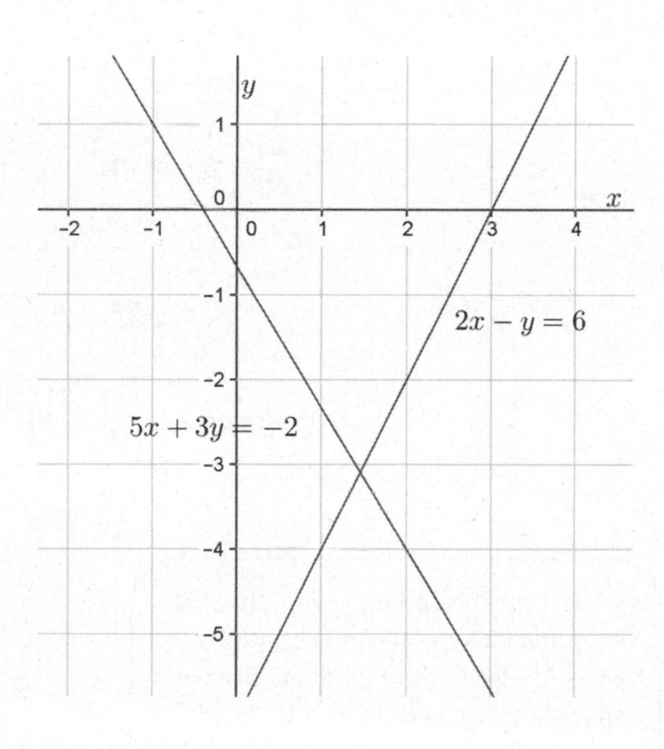

$$3(2x - y = 6)$$
$$6x - 3y = 18$$
$$\begin{cases} 5x + 3y = -2 \\ 6x - 3y = 18 \end{cases}$$
$$5x + 3y + 6x - 3y = -2 + 18$$
$$11x = 16$$
$$x = \frac{16}{11}$$

$$2\left(\frac{16}{11}\right) - y = 6$$
$$-y = \frac{34}{11}$$
$$y = -\frac{34}{11}$$

The solution is $\left(\frac{16}{11}, -\frac{34}{11}\right)$.

2. $\begin{cases} -6x - 2y = -3 \\ -8x + 2y = 7 \end{cases}$

I notice that since the first equation has $-2y$ and the second equation has $+2y$, when I add the equations together, y will be eliminated, and I can solve for x first.

$$-6x - 2y - 8x + 2y = -3 + 7$$
$$-14x = 4$$
$$x = -\frac{4}{14}$$

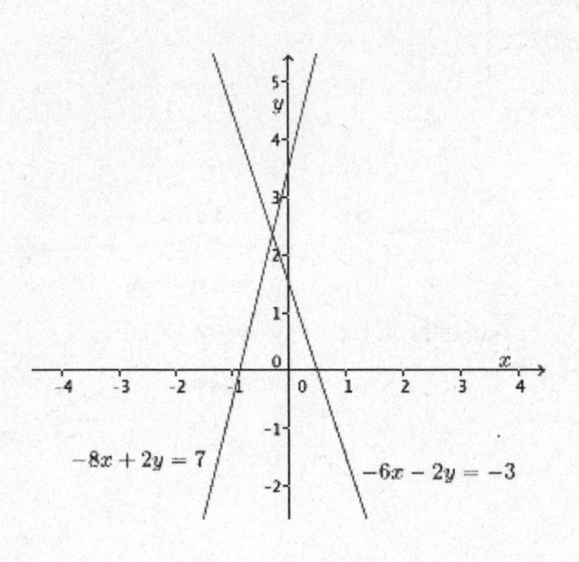

$$-6\left(-\frac{4}{14}\right) - 2y = -3$$
$$\frac{12}{7} - 2y = -3$$
$$-2y = -\frac{33}{7}$$
$$y = \frac{33}{14}$$

The solution is $\left(-\frac{4}{14}, \frac{33}{14}\right)$.

3. $\begin{cases} y = -2x + 7 \\ 6x + 3y = 21 \end{cases}$

When I substituted for y with the first equation into the second equation, $6x + 3(-2x + 7) = 21$, it resulted in an identity, namely $21 = 21$. This means the two equations will graph the same line.

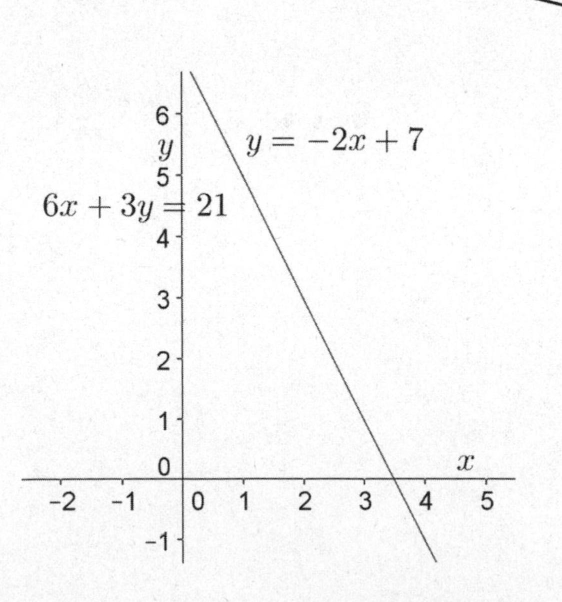

These equations define the same line. Therefore, this system will have infinitely many solutions.

4. $\begin{cases} -2x + 4y = 11 \\ x - 2y = 9 \end{cases}$

When I multiplied the second equation by 2, I eliminated both x and y. The result was an untrue statement, namely $0 \neq 29$. I should check the slopes and y-intercepts.

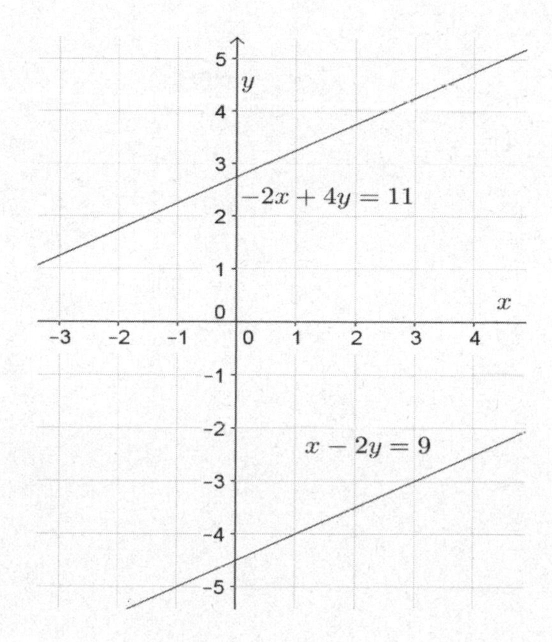

The equations graph as distinct lines. The slopes of these two equations are the same, and the y-intercepts are different, which means they graph as parallel lines. Therefore, this system will have no solution.

Determine the solution, if it exists, for each system of linear equations. Verify your solution on the coordinate plane.

1. $\begin{cases} \frac{1}{2}x + 5 = y \\ 2x + y = 1 \end{cases}$

2. $\begin{cases} 9x + 2y = 9 \\ -3x + y = 2 \end{cases}$

3. $\begin{cases} y = 2x - 2 \\ 2y = 4x - 4 \end{cases}$

4. $\begin{cases} 8x + 5y = 19 \\ -8x + y = -1 \end{cases}$

5. $\begin{cases} x + 3 = y \\ 3x + 4y = 7 \end{cases}$

6. $\begin{cases} y = 3x + 2 \\ 4y = 12 + 12x \end{cases}$

7. $\begin{cases} 4x - 3y = 16 \\ -2x + 4y = -2 \end{cases}$

8. $\begin{cases} 2x + 2y = 4 \\ 12 - 3x = 3y \end{cases}$

9. $\begin{cases} y = -2x + 6 \\ 3y = x - 3 \end{cases}$

10. $\begin{cases} y = 5x - 1 \\ 10x = 2y + 2 \end{cases}$

11. $\begin{cases} 3x - 5y = 17 \\ 6x + 5y = 10 \end{cases}$

12. $\begin{cases} y = \frac{4}{3}x - 9 \\ y = x + 3 \end{cases}$

13. $\begin{cases} 4x - 7y = 11 \\ x + 2y = 10 \end{cases}$

14. $\begin{cases} 21x + 14y = 7 \\ 12x + 8y = 16 \end{cases}$

Example 1

The sum of two numbers is 361, and the difference between the two numbers is 173. What are the two numbers?

Example 2

There are 356 eighth-grade students at Euclid's Middle School. Thirty-four more than four times the number of girls is equal to half the number of boys. How many boys are in eighth grade at Euclid's Middle School? How many girls?

Example 3

A family member has some five-dollar bills and one-dollar bills in her wallet. Altogether she has 18 bills and a total of \$62. How many of each bill does she have?

Example 4

A friend bought 2 boxes of pencils and 8 notebooks for school, and it cost him \$11. He went back to the store the same day to buy school supplies for his younger brother. He spent \$11.25 on 3 boxes of pencils and 5 notebooks. How much would 7 notebooks cost?

Exercises

1. A farm raises cows and chickens. The farmer has a total of 42 animals. One day he counts the legs of all of his animals and realizes he has a total of 114. How many cows does the farmer have? How many chickens?

2. The length of a rectangle is 4 times the width. The perimeter of the rectangle is 45 inches. What is the area of the rectangle?

3. The sum of the measures of angles x and y is $127°$. If the measure of $\angle x$ is $34°$ more than half the measure of $\angle y$, what is the measure of each angle?

Name ________________________________ Date ________________

1. Small boxes contain DVDs, and large boxes contain one gaming machine. Three boxes of gaming machines and a box of DVDs weigh 48 pounds. Three boxes of gaming machines and five boxes of DVDs weigh 72 pounds. How much does each box weigh?

2. A language arts test is worth 100 points. There is a total of 26 questions. There are spelling word questions that are worth 2 points each and vocabulary word questions worth 5 points each. How many of each type of question are there?

1. Two numbers have a sum of 853 and a difference of 229. What are the two numbers?

Let x represent one number and y represent the other number.

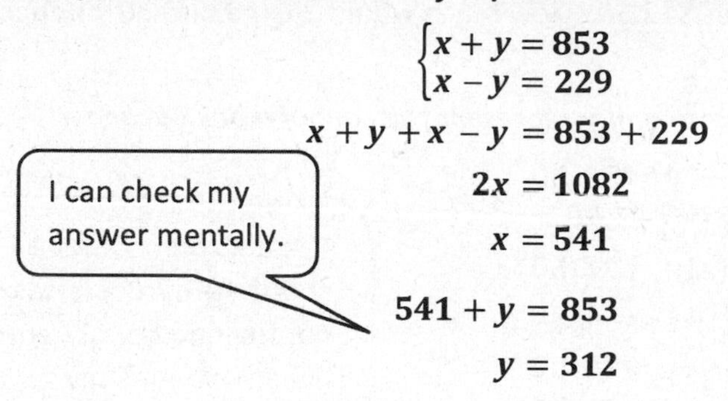

$$\begin{cases} x + y = 853 \\ x - y = 229 \end{cases}$$

$$x + y + x - y = 853 + 229$$

$$2x = 1082$$

$$x = 541$$

$$541 + y = 853$$

$$y = 312$$

> Sum means I add the two numbers together, and difference means I subtract one number from the other. Since I don't know either number, I need to define my variables with two different letters.

The solution is $(541, 312)$. The two numbers are 541 and 312.

2. The sum of the ages of two sisters is 36. The younger sister is 6 more than a fifth of the older sister's age. How old is each sister?

Let x represent the age of the younger sister and y represent the age of the older sister.

> A fifth of the older sister's age means to multiply $\frac{1}{5}$ times y, the age of the older sister.

$$\begin{cases} x + y = 36 \\ x = 6 + \frac{1}{5}y \end{cases}$$

> I will use the substitution method since x is isolated. I will replace x with $6 + \frac{1}{5}y$ in the first equation.

$$6 + \frac{1}{5}y + y = 36$$

$$6 + \frac{6}{5}y = 36$$

$$\frac{6}{5}y = 30$$

$$y = 25$$

$$x + 25 = 36$$

$$x = 11$$

> If I let x represent the age of the older sister and y represent the age of the younger sister, the second equation would be $y = 6 + \frac{1}{5}x$. I would use the same method to solve.

Check:

$$11 = 6 + \frac{1}{5}(25)$$

$$11 = 6 + 5$$

$$11 = 11$$

The solution is $(11, 25)$. The older sister is 25 years old, and the younger sister is 11 years old.

3. Some friends went to the local movie theater and bought three buckets of large popcorn and four boxes of candy. The total for the snacks was $30.50. The last time you were at the theater, you bought a large popcorn and two boxes of candy, and the total was $12.50. How much would 2 large buckets of popcorn and 3 boxes of candy cost?

Let x represent the cost of a large bucket of popcorn and y represent the cost of a box of candy.

I have choices. I could eliminate x by multiplying the second equation by -3 or eliminate y by multiplying the second equation by -2.

The question is asking about the *cost* of items, not the number of items. I need to define my variables as the *cost* of each item.

$$\begin{cases} 3x + 4y = 30.50 \\ x + 2y = 12.50 \end{cases}$$

$$-2(x + 2y = 12.50)$$

$$-2x - 4y = -25$$

$$\begin{cases} 3x + 4y = 30.50 \\ -2x - 4y = -25 \end{cases}$$

$$3x + 4y - 2x - 4y = 30.50 - 25$$

$$3x - 2x = 5.50$$

$$x = 5.50$$

$$5.50 + 2y = 12.50$$

$$2y = 7$$

$$y = 3.50$$

***The solution is* $(5.50, 3.50)$.**

Once I find out the cost of each item, I can determine the cost for 2 large popcorns and 3 boxes of candy.

Check:

$$3(5.50) + 4(3.50) = 30.50$$

$$16.50 + 14 = 30.50$$

$$30.50 = 30.50$$

Since a large bucket of popcorn costs* $\$5.50$ *and a box of candy costs* $\$3.50$*, then the equation to find the cost of two large buckets of popcorn and three boxes of candy is* $2(5.50) + 3(3.50) = 11 + 10.50$*, which is equal to* 21.50*. Therefore, the cost of two large buckets of popcorn and three boxes of candy is* $\$21.50$*.

1. Two numbers have a sum of 1,212 and a difference of 518. What are the two numbers?

2. The sum of the ages of two brothers is 46. The younger brother is 10 more than a third of the older brother's age. How old is the younger brother?

3. One angle measures 54 more degrees than 3 times another angle. The angles are supplementary. What are their measures?

4. Some friends went to the local movie theater and bought four large buckets of popcorn and six boxes of candy. The total for the snacks was $46.50. The last time you were at the theater, you bought a large bucket of popcorn and a box of candy, and the total was $9.75. How much would 2 large buckets of popcorn and 3 boxes of candy cost?

5. You have 59 total coins for a total of $12.05. You only have quarters and dimes. How many of each coin do you have?

6. A piece of string is 112 inches long. Isabel wants to cut it into 2 pieces so that one piece is three times as long as the other. How long is each piece?

Mathematical Modeling Exercise

(1) If t is a number, what is the degree in Fahrenheit that corresponds to t°C?

(2) If t is a number, what is the degree in Fahrenheit that corresponds to $(-t)$°C?

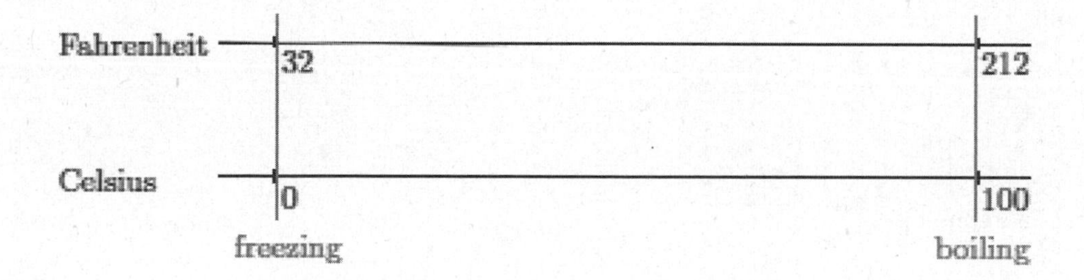

Exercises

Determine the corresponding Fahrenheit temperature for the given Celsius temperatures in Exercises 1–5.

1. How many degrees Fahrenheit is $25°\text{C}$?

2. How many degrees Fahrenheit is $42°\text{C}$?

3. How many degrees Fahrenheit is $94°\text{C}$?

4. How many degrees Fahrenheit is $63°\text{C}$?

5. How many degrees Fahrenheit is $t°\text{C}$?

Name____ Date ____________________

Use the equation developed in class to answer the following questions:

1. How many degrees Fahrenheit is $11°C$?

2. How many degrees Fahrenheit is $-3°C$?

3. Graph the equation developed in class, and use it to confirm your results from Problems 1 and 2.

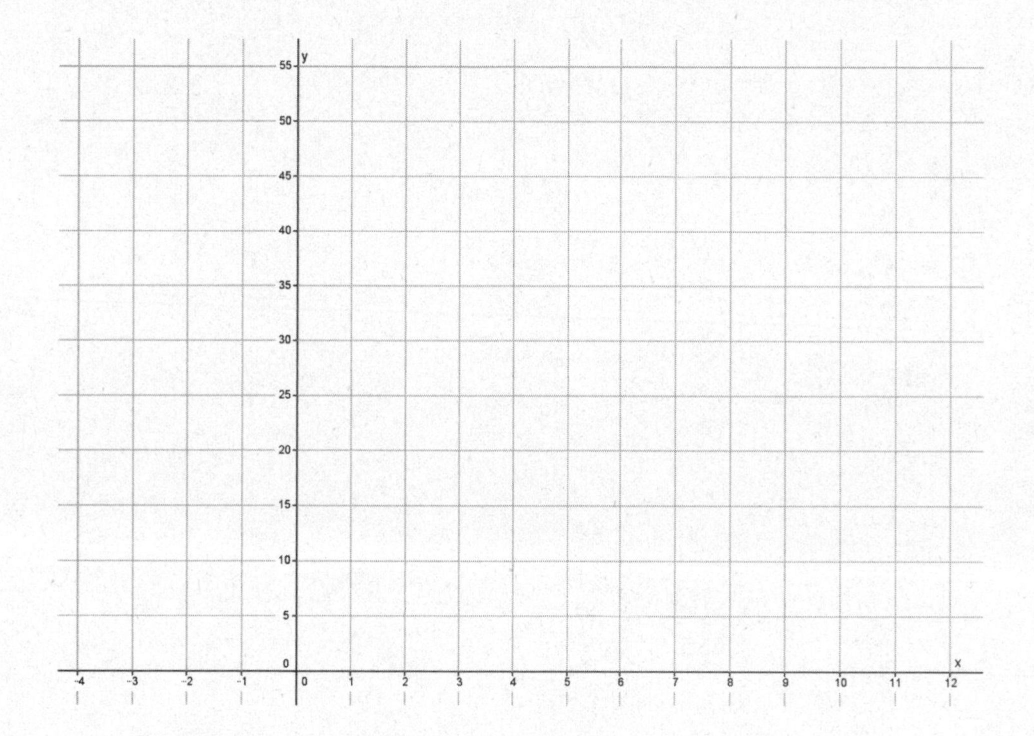

1. Does the equation $t°\mathrm{C} = (32 + 1.8t)°\mathrm{F}$ work for any rational number t? Check that it does with $t = 12\frac{1}{5}$ and $t = -12\frac{1}{5}$.

I will use substitution with $t = 12\frac{1}{5}$ and $t = -12\frac{1}{5}$.

$$\left(12\frac{1}{5}\right)°C = \left(32 + 1.8 \times 12\frac{1}{5}\right)°F = (32 + 21.96)°F = 53.96\ °F$$

This means that $12\frac{1}{5}$ °C is the same as 53.96°F.

$$\left(-12\frac{1}{5}\right)°C = \left(32 + 1.8 \times \left(-12\frac{1}{5}\right)\right)°F = (32 - 21.96)°F = 10.04°F$$

2. Knowing that $t°\mathrm{C} = \left(32 + \frac{9}{5}t\right)°\mathrm{F}$ for any rational number t, show that for any rational number d, $d°\mathrm{F} = \left(\frac{5}{9}(d - 32)\right)°\mathrm{C}$.

I will write down everything I know from the problem and lesson.

From the lesson, I know that $d°\mathrm{F} = \left(32 + \frac{9}{5}t\right)°\mathrm{F}$.

That implies that $d = \left(32 + \frac{9}{5}t\right)$.

From the problem, I know that $t°\mathrm{C} = \left(32 + \frac{9}{5}t\right)°\mathrm{F}$.

From the lesson, I know that $t°\mathrm{C} = d°\mathrm{F}$.

I will use these equations to help me show that $d°\mathrm{F} = \left(\frac{5}{9}(d - 32)\right)°\mathrm{C}$.

I will start by solving for t.

Since $d°F$ can be found by $\left(32+\frac{9}{5}t\right)°F$, then $d=\left(32+\frac{9}{5}t\right)$, and $d°F=t°C$. Substituting $d=\left(32+\frac{9}{5}t\right)$ into $d°F$ we get

$$d°F=\left(32+\frac{9}{5}t\right)°F$$
$$d=32+\frac{9}{5}t$$
$$d-32=\frac{9}{5}t$$
$$\frac{5}{9}(d-32)=t$$

Now that we know $t=\frac{5}{9}(d-32)$, then $d°F=\left(\frac{5}{9}(d-32)\right)°C$.

Once I know t, I can substitute into $t°\mathrm{C}=d°\mathrm{F}$ to show that for any rational number d, $d°\mathrm{F}=\left(\frac{5}{9}(d-32)\right)°\mathrm{C}$.

1. Does the equation $t°\text{C} = (32 + 1.8t)°\text{F}$ work for any rational number t? Check that it does with $t = 8\frac{2}{3}$ and $t = -8\frac{2}{3}$.

2. Knowing that $t°\text{C} = \left(32 + \frac{9}{5}t\right)°\text{F}$ for any rational number t, show that for any rational number d, $d°\text{F} = \left(\frac{5}{9}(d - 32)\right)°\text{C}$.

3. Drake was trying to write an equation to help him predict the cost of his monthly phone bill. He is charged $35 just for having a phone, and his only additional expense comes from the number of texts that he sends. He is charged $0.05 for each text. Help Drake out by completing parts (a)–(f).

   a. How much was his phone bill in July when he sent 750 texts?

   b. How much was his phone bill in August when he sent 823 texts?

   c. How much was his phone bill in September when he sent 579 texts?

   d. Let y represent the total cost of Drake's phone bill. Write an equation that represents the total cost of his phone bill in October if he sends t texts.

   e. Another phone plan charges $20 for having a phone and $0.10 per text. Let y represent the total cost of the phone bill for sending t texts. Write an equation to represent his total bill.

   f. Write your equations in parts (d) and (e) as a system of linear equations, and solve. Interpret the meaning of the solution in terms of the phone bill.

Exercises

1. Identify two Pythagorean triples using the known triple $3, 4, 5$ (other than $6, 8, 10$).

2. Identify two Pythagorean triples using the known triple $5,12, 13$.

3. Identify two triples using either $3, 4, 5$ or $5, 12, 13$.

Use the system $\begin{cases} x + y = \frac{t}{s} \\ x - y = \frac{s}{t} \end{cases}$ to find Pythagorean triples for the given values of s and t. Recall that the solution in the form of $\left(\frac{c}{b}, \frac{a}{b}\right)$ is the triple a, b, c.

4. $s = 4, t = 5$

5. $s = 7, t = 10$

6. $s = 1, t = 4$

7. Use a calculator to verify that you found a Pythagorean triple in each of the Exercises 4–6. Show your work below.

Lesson Summary

A Pythagorean triple is a set of three positive integers that satisfies the equation $a^2 + b^2 = c^2$.

An infinite number of Pythagorean triples can be found by multiplying the numbers of a known triple by a whole number. For example, 3, 4, 5 is a Pythagorean triple. Multiply each number by 7, and then you have 21, 28, 35, which is also a Pythagorean triple.

The system of linear equations, $\begin{cases} x + y = \frac{t}{s} \\ x - y = \frac{s}{t} \end{cases}$, can be used to find Pythagorean triples, just like the Babylonians did 4,000 years ago

Name ________________________________ Date ________________

Use a calculator to complete Problems 1–3.

1. Is $7, 20, 21$ a Pythagorean triple? Is $1, \frac{15}{8}, \frac{17}{8}$ a Pythagorean triple? Explain.

2. Identify two Pythagorean triples using the known triple $9, 40, 41$.

3. Use the system $\begin{cases} x + y = \frac{t}{s} \\ x - y = \frac{s}{t} \end{cases}$ to find Pythagorean triples for the given values of $s = 2$ and $t = 3$. Recall that the solution in the form of $\left(\frac{c}{b}, \frac{a}{b}\right)$ is the triple a, b, c. Verify your results.

Lesson Notes

Any three numbers, a, b, and c, that satisfy $a^2 + b^2 = c^2$ are considered a triple. A Pythagorean triple is a set of three *whole numbers*, a, b, and c, that satisfy the equation $a^2 + b^2 = c^2$.

Examples

1. Identify a Pythagorean triple (numbers that satisfy $a^2 + b^2 = c^2$), using the known Pythagorean triple 5, 12, 13.

   I need to multiply the known triple by a whole number to ensure that I produce a Pythagorean triple.

   Answers will vary.

   A triple is $10, 24, 26$. I found these by multiplying each of $5, 12$, and 13 by 2.

2. Identify a triple (numbers that satisfy $a^2 + b^2 = c^2$), using the known Pythagorean triple 5, 12, 13.

   To produce a triple, I will multiply each number in the known triple by a number between 0 and 1.

   Answers will vary.

   A triple is $3.5, 8.4, 9.1$. I found these by multiplying each of 5, 12, and 13 by 0.7.

3. Use the system $\begin{cases} x + y = \frac{t}{s} \\ x - y = \frac{s}{t} \end{cases}$ to find Pythagorean triples for the given values of s and t. Recall that the solution, in the form of $\left(\frac{c}{b}, \frac{a}{b}\right)$, is the triple, a, b, c.

$s = 2, t = 5$

This system will produce triples only if $t > s$.

$$\begin{cases} x + y = \frac{5}{2} \\ x - y = \frac{2}{5} \end{cases}$$

I will use elimination to solve the system by summing two equations.

$$x + y + x - y = \frac{5}{2} + \frac{2}{5}$$

$$2x = \frac{29}{10}$$

$$x = \frac{29}{20}$$

$$\frac{29}{20} + y = \frac{5}{2}$$

$$y = \frac{5}{2} - \frac{29}{20}$$

$$y = \frac{21}{20}$$

Then the solution is $\left(\frac{29}{20}, \frac{21}{20}\right)$***, and the triple is*** $\mathbf{20, 21, 29}$.

I write the numerators and denominator in ascending order. The denominator of both equations is b, the smaller numerator is a, and the larger numerator is c.

4. Use a calculator to verify that you found a Pythagorean triple in Problem 2. Show your work below.

For the triple $\mathbf{20, 21, 29}$:

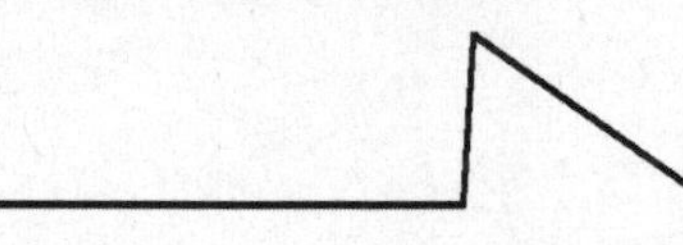

$$20^2 + 21^2 = 29^2$$

$$400 + 441 = 841$$

$$841 = 841$$

The longest side of the right triangle is the hypotenuse identified by c.

If the triple I found in Problem 2 cannot be verified, I will need to go back and check my work.

1. Explain in terms of similar triangles why it is that when you multiply the known Pythagorean triple $3, 4, 5$ by 12, it generates a Pythagorean triple.

2. Identify three Pythagorean triples using the known triple $8, 15, 17$.

3. Identify three triples (numbers that satisfy $a^2 + b^2 = c^2$, but a, b, c are not whole numbers) using the triple $8, 15, 17$.

Use the system $\begin{cases} x + y = \frac{t}{s} \\ x - y = \frac{s}{t} \end{cases}$ to find Pythagorean triples for the given values of s and t. Recall that the solution, in the form of $\left(\frac{c}{b}, \frac{a}{b}\right)$, is the triple a, b, c.

4. $s = 2, t = 9$

5. $s = 6, t = 7$

6. $s = 3, t = 4$

7. Use a calculator to verify that you found a Pythagorean triple in each of the Problems 4–6. Show your work.

Credits

Great Minds® has made every effort to obtain permission for the reprinting of all copyrighted material. If any owner of copyrighted material is not acknowledged herein, please contact Great Minds for proper acknowledgment in all future editions and reprints of this module.